The Unfinished Revolution

MICHAEL L. DERTOUZOS

THE Unfinished Revolution

HUMAN-CENTERED COMPUTERS AND WHAT THEY CAN DO FOR US

HarperCollins*Publishers*

HarperCollins books may be purchased for educational, business, or sales promotional use. For information please write: Special Markets Department, HarperCollins Publishers, Inc., 10 East 53rd Street, New York, NY 10022.

FIRST EDITION

Designed by Joy O'Meara

Library of Congress Cataloging-in-Publication Data
Dertouzos, Michael L.
    Unfinished revolution : human-centered computers and what they can do for us / Michael L. Dertouzos.
      p. cm.
    ISBN 0-06-662067-8
      1. Human-computer interaction. 2. User interfaces (Computer systems) I. Title.

QA76.9.H85 D46 2001
004'.01'9—dc21

                                                                            00-059767

01 02 03 04 05 RRD 10 9 8 7 6 5 4 3 2 1

To Kiera

Contents

Preface

The Unfinished Revolution sets forth a radically new direction for information technology, and the way it could be used to make computer systems serve people . . . rather than the other way around.

Fundamental change is overdue. As individuals and organizations everywhere scramble to take advantage of the Web, the Internet, and a myriad of new gadgets, they want to know what they should do. The media, vendors, and pundits respond with advice, trends, possibilities, and opinions in the thousands. Yet the overwhelming outcome of this frenzy is a feeling of profound confusion by ordinary users and specialists alike.

The confusion is justified. Does all this new and exciting technology make us "better off"? Or are we headed toward greater complexity, increased frustration, and a human burden that will grow in proportion to the gadgets and programs that surround us? We certainly can be better off with information technology. But not the way we are headed. Without a fundamentally new approach to computing, the confusion will get worse and the Information Revolution will remain unfinished.

The new approach has been taking shape in my head for more than a decade, although it didn't gel with a name and an action plan until recently. It started with the frustration that I and others felt as we repeatedly tried to harness computers to our purposes, only to discover that we were the ones who ended up under the yoke. The idea became stronger as the complexities of computers increased, as the features of programs that no one needed multiplied, and as people became increasingly trapped in the use of systems that pretended to change while remaining stagnant and distant from human purposes.

I have called the new approach human-centric computing, and the machines human-centered, to emphasize that from now on, computer systems should focus on our needs and capabilities, instead of forcing us to bow down to their complex, incomprehensible, and mechanistic details. Human-centered computers are not a fantasy. They can be built, right now, with current and emerging technologies. We can even begin with the computers we already have, merely by changing the way we use them. This book lays out the human-centric approach by explaining in everyday language the five basic forces that define it, the ways people will use it, and the impact it could have upon our lives.

What the book does not do is assemble a collage of futuristic vignettes designed to impress through shock. Such scenarios, easy to concoct, are less exciting than what is really likely to happen. Neither do I rehash the faddish mantras to make computers more "intelligent," or more "user-friendly." These are mere restatements of our wish to get out of the mess we are in. They do not show the way! Predicting the future is difficult, but the odds get better when you are trying to build it, rather than guess it. This is the approach that served me well 20 years ago in forecasting the Information Marketplace that is rising fast among us. And it is the approach I am taking now, together with my colleagues at MIT, as we engage in the ambitious pursuit of human-centric systems. It is also the approach increasingly taken by other cutting-edge research institutions and companies around the world as they explore and craft their own visions of the future.

The contributions of many people have influenced my thinking. I am grateful to them and especially to my colleagues at the MIT Laboratory for Computer Science. But I do not speak for them and they should not be held responsible for what I say here. *The Unfinished Revolution* is a declaration of my personal ideas, passions, and beliefs about human-centered systems.

I wrote this book for people who use computers, and for the technologists who build them, to offer a new insight about where we should steer the computing juggernaut. I hope the book sets forth a new philosophy for information technology, and provides a manifesto for turning it into reality. I hope it inspires computer users and builders to fuel the torch of human-centric computing with their creative ideas. And I hope it sparks a revolution within the computer revolution.

It's high time we did so!

Michael L. Dertouzos
Weston, Massachusetts

Acknowledgments

I thank my colleagues Anant Agarwal, Rodney Brooks, Frans Kaashoek, and Victor Zue for working together with me to shape the vision of a new breed of human-centered computer systems.

I thank, too, the following people who reviewed all or parts of the manuscript and made helpful suggestions: Hal Abelson, Mark Ackerman, John Ankcorn, Hari Balakrishnan, Rodney Brooks, Dave Clark, Srini Devadas, Hamish Fraser, Stephen Garland, Eric Grimson, John Guttag, David Karger, Philip Khoury, Isaac Kohane, Narayana Murthy, Ron Rivest, Gerald Ruderman, Ed Rykken, Howie Shrobe, David Siegel, Lynn Andrea Stein, Ralph Swick, David Tennenhouse, Paul Viola, Steve Ward, and Victor Zue.

I also thank my assistant Anne Wailes for helping in so many ways; communications expert Patti Richards for helping get the book out there; my editor John Benditt at *Technology Review,* where I have written several columns from which I've borrowed some of my own words; my agent Ike Williams for his unbridled enthusiasm and support; and Adrian Zackheim, who found the time while running Harper Information books to also act as the energetic and thoughtful editor of *The Unfinished Revolution.*

Freelance editor and writer Mark Fischetti has my deepest thanks for patiently and creatively working with me to structure language, ideas, and approach for the entire book.

Finally, I am grateful to my wife, Catherine Liddell Dertouzos, for being enthusiastic about this project, for contributing her ideas, and for supporting me during the writing.

One
WHY CHANGE

Weird animals surround me in my home, at work, everywhere I go. Every day I must spend hours feeding them, healing them, waiting for them. And the fighting! They hold each other hostage in asphyxiating head-locks. I scream at them, but they just grunt or stare back stupidly. When we do get along, and I'm feeling affection for them, they suddenly turn around and bite a chunk off my hide.

You are surrounded by these creatures, too—the personal computers, laptops, handheld assistants, printers, Internet-savvy phones, music storage drives, and other digital wonders. They are everywhere and multiplying fast. Yet instead of serving us, we are serving them. We wait endlessly for our computers to boot up, and for bulky Web pages to paint themselves on our screens. We stand perplexed in front of incomprehensible system messages, and wait in frustration on the phone for computerized assistance. We constantly add software upgrades, enter odd instructions, fix glitches, only to sit in maddening silence when our machines crash, forcing us to start all over again, hoping against hope that they didn't take a piece of our intellectual hide with them. We'd never live in a house, work in an office, or ride in a car where we had to put up with a menagerie of such beasts. Yet we do it every day with our computer menagerie.

We shouldn't have to.

We have already gone so far down the road of serving computers that we've come to accept our servitude as necessary. It isn't. It is time for us to rise up with a profound demand: "Make our computers simpler to use!" Make them talk with us, do things for us, get the information we want, help us work with other people, and adapt to our individual needs. Only then will computers make us productive and truly serve us, instead of the other way around.

Is this possible? Certainly.

Before I reveal an entirely new approach to computer systems and their uses—a new plan for *human-centric computing*—let me assure you that in our new century, we have every right to expect fundamental reform. For 40 years computers have been shrines to which we pay dutiful homage. When something goes wrong, the "user"— you and I—feel that if we somehow had behaved better the trouble would not have arisen. But we are not at fault. The trouble lies in the current approach to computing.

If computers are to live up to the promise of serving us, they will have to change drastically and never again subject us to the frustrating experiences we have all shared.

Several colleagues from the MIT Laboratory for Computer Science and I are flying to Taiwan. I have been trying for three hours to make my new laptop work with one of these "smart cards" that plug into the machine and download my personal calendar. When the card software is happy, the operating system complains, and vice versa. Irritated, I turn to Tim Berners-Lee, sitting next to me, who graciously offers to assist. After an hour the inventor of the Web admits that the task is beyond his capabilities. I turn to Ron Rivest, inventor of RSA public key cryptography, and ask him to help. He declines, exhibiting his wisdom. A young faculty member behind us speaks up: "You guys are too old. Let me do it." He gives up after an hour and a half. So I go back to my "expert" approach of typing random entries into the various wizards and lizards that keep popping up on the screen. After two more hours, and two batteries, I make it work, by sheer accident and without remembering how.

My friends on this flight were hardly incompetent. The problem was what I call the "unintegrated systems fault." Technologists design today's hardware and software systems without worrying enough about how these different pieces will work together. If the slightest conflict arises among an operating system, a communications network, a digital camera, a printer, or any other device, the modules become deadlocked, as do their makers, who point to one another, leaving you to resolve their differences. After I published this Taiwan anecdote in an August 1999 article in *Scientific American,* I received scores of letters from people who said, "I know exactly what you are talking about. Please fix it." The problem is not simply a "bug" to be worked out in existing systems, but rather an endemic mind-set that has characterized computer design for decades. Only a radical change can fix it.

It's 11 P.M. and I check my e-mail. Ninety-eight new messages have arrived since yesterday. At 2 to 3 minutes per message, my average response time, I'll need 4 hours to handle them. I'd like to grant them my highest security classification, DBR—"destroy before reading."

How do we handle this "overload fault?" We don't. Mostly, we feel guilty if we cannot respond to all the messages that come our way. Better e-mail software can relieve a lot of this burden. Better human behavior can go further. Human-centric computing means more than changing the hardware and software of computer systems. We must also improve the ways we use technology.

My son is searching the Web for information on Vespas, the Italian scooters that conquered Europe in the 1950s, which he loves to restore. The search engine has given him 2,545 hits and he is busy checking them out. His eyes squint and his brain labors to minimize the time he needs to decide whether he should keep or toss each entry. I imagine him in an ancient badlands, furiously shoveling through 2,545 mountains of dirt, looking for one nugget of hidden treasure. His shovel is diamond studded and it is stamped "high tech," so he is duly modern. Yet he is still shoveling!

There are two problems here. First, the "manual labor fault," which reflects the lack of automation on today's Web and in all of today's computer systems. We do not yet off-load human brain work and eyeball work onto our machines. We shovel and shovel, doing by ourselves mental labor we shouldn't have to do. The second problem is the "information access fault," which reflects our inability to get at the information we need when we need it. Both faults can be repaired.

The automatic answering system greets you with its murderous "You have reached the Tough Luck Corporation. If you want Marketing, press 1. If you want Engineering, press 2. If you want a company directory, press 3, then enter the letters of the last name of the person you wish to speak to . . ."

Here we have a human being, on whose head a price cannot be set, obediently executing instructions dispensed by a $100 computer. Welcome to the "human servitude fault." *You* are serving the inhuman machine, and its inhuman owners who got away saving a few dollars of operator time by squandering valuable pieces of your life and that of millions of other people. What glory: The highest technology artifacts in the world have become our masters, reintroducing us to human slavery more than a century after its abolition. Our docility in putting up with this abuse is reprehensible.

Then there is the famous "crash fault." You are working along nicely and something untoward happens in the bowels of the machine, causing it to crash. If you haven't done your "duty" of saving your work every few minutes, you are in for some grief. And when you reboot the system, you are rewarded for your tolerance of the crash by a reprimand implying that you turned off your machine improperly! Things don't have to be that way. Telephone switching systems hardly ever crash, yet they use software in their computers that is just as complex as the software in your PC.

More faults like these abound: the "excessive learning fault," where a word processing program, which does what a pencil used to do, only somewhat differently, comes with a 600-page manual. The "feature overload fault," where megabytes of software features you'll never use are stuffed into your machine, making the features you do want to use hard to find, slow, and prone to crashes. The "fake intelligence fault," where the machine purports to be intelligent but is not, getting in your way instead of helping you. The "waiting fault," where we ask the machine to turn on or off, and we wait unbearable minutes until it condescends to do so. The "ratchet fault," where layers and layers of software piled up on top of each other through the ages create a spaghetti-like mess that even its maker can't untangle.

It gets worse. Trendy handheld PDAs (personal digital assistants) demand that you learn entire new sets of commands, and go back to first grade to relearn how your fingers should pen letters of the alphabet, when you write on their little screens, so their programs can understand you. These quirky devices overlap each other's functions and pose more demands on our attention. "Where should I put my calendar—in my PC, my PDA, or my brand new cell phone?" "Ah!" bellows the voice of an all-knowing friend! "Get a synchronizer to keep them all in step." And so it is that yet another piece of software enters your life, with its thick manual, new commands, and many versions yet to come.

Periodically, on top of all these insults, the dreaded time arrives when you must change computers. Suddenly, all your work is hanging by a thread. You will squander entire days trying to reinject your old programs and files onto the new machine. And once you have ensured their survival, a whole bunch of new software conflicts will rear their ugly heads.

So total is our brainwashing and habitual acceptance of these indignities that even as they are happening, we brag that we have the latest breed of this or that machine that runs 30 percent faster than our neighbor's computer and has this new set of great features!

We need a radical change.

It may sound harsh, but even though they have helped us do amazing things we never could have done without them, computers have increased hype more than productivity. The marketers call them user-friendly, knowing that they are difficult to use. And despite the chest beating about the Internet giving a voice to people throughout the world, the new technology is only used by a tiny fraction of the human population.

The real utility of computers, and the true value of the Information Revolution, still lie ahead. And I'm not talking about a few improvements. The Web and the Internet of today, compared to where we are headed, are like steam engines compared to the modern industrial world. By the time information systems reach jet-plane status, well into this century, we will focus on utility over fads, triple our productivity, use our computers as naturally and easily and with as much pleasure as we now use our cars and refrigerators, and hear the voices of hundreds of millions more people—if we abandon our self-defeating path toward unbridled and growing machine complexity.

We must set a new goal which is as obvious and simple as it is powerful: *Information technology should help people do more by doing less.* Human-centric computing is the approach that leads to that goal. It is what will finish the Unfinished Revolution.

Charting New Terrain

If the quirky machines that surround you are causing you grief, imagine the mess you'll be in when there are 10 times as many of these creatures biting at you in the next few years. That's where we are headed with the huge variety of new devices coming our way. Let's not be passive victims. Let's grab "progress" by the throat and redirect it so it serves us. If we don't deliberately do so, starting now, tomorrow's much larger menagerie of hardware and software systems will make our lives even more servile and complicated.

To achieve human-centric computing, we must pay attention to both the human and the computer side of the relationship. We begin with the raw material we have to work with—the computing terrain.

We must understand how it is shaped and think ahead about how it will change, because technology constantly evolves.

By 2015, single-processor PCs will level out at around 50 times the speed of Year 2000 machines, because of fundamental limits on the smallest circuits that can be "printed" on a chip. To get greater performance, designers will harness microprocessors together, like horses on a cart, up to a thousand or so, before the tiny machines get in each other's way. These combined moves will make future machines tens of thousands of times faster. During the same period, the average communication speed between machines on the Internet will increase a few hundred times, using today's pipes—mostly telephone lines and television's coaxial cable. The number of people who use wireless communications will grow dramatically, but communication speed will remain well below that of future wire line phones and coax cables. The speed of communication among stationary machines will get another thousandfold boost when sometime in the next two decades the trillion-dollar plunge is taken, as it inevitably will be, by telephone, cable TV, and other companies to thread every home and office in the industrial world with glass fiber lines. Ultimately, these high-speed terrestrial links will interconnect a huge number of antennas that will define increasingly smaller wireless "cells," with ever-increasing bandwidth for roving users. As a result, communication speeds over the Internet will eventually become ten thousand times faster than what they are today. Storage capacity on computers will keep up with this maniacal pace and costs will continue to drop from the Year 2000 level of $15 for a gigabyte (the equivalent of 500 paperbacks) to well below $1.

"Who needs all these gigas of power, speed, and storage?" you may ask. You do! But you don't know it, because the numbers measure what machines do, rather than what people care about. You want to know: "How quickly can tomorrow's system locate and ship to me a replacement part for my bathroom fixture? How well can I collaborate on producing a manual for my company's new product with a coworker who lives eight time zones away? Can I tell the computer to book me a flight to Israel, and have it carry out all the negotiations?"

A human-centric computer that can perform these functions quickly and effectively with minimal instruction from you will have to be simpler to use on the outside, which means it will have to become more complex on the inside—hence the need for all the gigas. But like a car, even though its inner workings will be complicated and powerful, all you should need to use it fully is the equivalent of a simple gas pedal, brake, and steering wheel. This ascent toward true human utility will take time, but we can accelerate the process if we are not lulled by the siren song of the gigas. Starting now, we must judge computers' performance by how well they satisfy our needs, not by how fast they spin their wheels.

While the "horsepower" of computers and communications will increase remarkably, three shifts in the new terrain will drive even greater change: interconnection of a growing number of appliances and physical devices to our computers; an increasing use of mobile computers connected through wireless communications; and a new breed of highly mobile software.

For the half century of their existence, essentially all the machines we have used have been operated by us. This is about to change in a big way. Run-of-the-mill appliances will become first-class computer citizens. Microprocessors with the ability to communicate with our computers throughout a house, an office building, or across the Internet will be embedded in lots of physical objects we care about. These interconnected appliances will weigh us as we stand in the bathroom in the morning; prepare most of our breakfast and have it ready just as we enter the kitchen (while also ordering foodstuffs that are depleted); deliver, as we eat, an urgent memo we have been anxiously awaiting and return our spoken reply; open the garage and lower the house temperature as we leave for the office; and announce, as we exit the driveway, a special morning program we were expecting and can now listen to while commuting directions are displayed on our windshield for avoiding the latest traffic jams. The processors will control the physical appliances tirelessly, 24 hours a day, while giving us instant access to them and the information on all our personal systems, and on the Web, when we want it, wherever we are, and on whatever device happens to be handy.

Putting microcomputers in physical devices isn't new. What's new is the promise that these physical appliances will be harnessed by tomorrow's computers to serve your needs. Many languages and systems are already being developed to help appliances communicate with computers. Unfortunately, they are following the patterns of today's computer and communication systems: They work, but they are complex and hard to use.

Interconnecting appliances to our computers is driven by a fundamental, natural force: Every day you interact with all sorts of physical things to achieve your purposes. Computerizing some of these exchanges so they become easier, faster, more reliable, and automatic will greatly enhance your ability to do more by doing less. This means that the number of these computerized appliances will far exceed the number of PCs. It also means that we'll computerize only those appliances whose utility justifies their interconnection "cost"—not everything in sight, as the hype suggests. If your main goal is to feed your family, you won't buy shoes with embedded chips. And even if you are rich, you may elect to sink your hands in your garden dirt, rather than use your computerized soil sensor.

The second big change in the new terrain is growth of wireless mobility. Tomorrow's computers, phones, and many other devices will be able to communicate easily without wires—be it across the room to control your entertainment center, across town to check on your house's security, or across a continent to help you reach an associate. By 2003 cellular phones, many with Internet access capability, will exceed in number the familiar wired telephones, which already exceed the number of PCs. The desire for wireless mobility is huge, for it stems from a powerful, natural force: People move. If machines can help us reach the humans and things we care about, wherever we and they may be, we can do even more by doing less. This change has already begun with laptops, PDAs, and cellular phones. But the rapidly advancing wireless terrain will extend it further, with the result that roving humans will get increasingly closer to the computer and communication power they now have at their desk.

The family reunion is about to begin at the island of Crete. Kiera arrives first and is shocked to find out that the hotel where they had planned to meet had a fire the night before. The smell of smoke is overwhelming. She pulls out her portable and says, "Urgent. Get me the relatives." Within one minute, 12 of the 15 relatives have joined the conversation. Three, already nearby, use their portables as two-way radios. Seven others are near high-speed networks and use their portables as network nodes. Of the remaining 6, 2 are in their cars heading for the hotel and 4 haven't left home and become connected through their home machines. All are alerted about the crisis. Michael, who is at home, volunteers to lead the search for alternatives. With the others listening in he says, "Get me Omni travel." The travel agent appears on Michael's full-size screen and is as upset as the travelers—15 commissions are at stake. She checks her hotel database and 30 seconds later sends to all the listeners a map showing three nearby hotels that have available rooms and meet the comfort level she knows they are after. Joan, who sees the map on her portable's screen, says she is passing near one of the hotels; she reports that it looks really good and has a majestic view. The travel agent connects to the hotel's machine and confirms that it can accommodate her clients and their needs. She books the new rooms, cuts her commission by 30 percent for good measure, and signs off. Twelve minutes after the discovery of the problem, the crisis has been averted and all the relatives have been redirected to the new location.

The third big change in the computing terrain will be in software. The devices we'll carry as we move will require software that can provide us with a "continuity" of services, regardless of which device we use. This will cause the software to become detached from specific devices and flow among them, carrying the functions we need, where and when we need them. For example, information about your health, diet, and caloric intake isn't nearly as useful on your office PC as it would be on your kitchen table's info outlet, or on your PDA when you're in a restaurant 5,000 miles from home and the crème brûlée appears on the dessert trolley. And when your daughter, sitting next to you in the kitchen, is dying to find out if she has an e-mail message from her boyfriend, she should be able to do so on the same device you used a second earlier to consult your diet plan.

This notion of dressing different machines with the information you want, where and when you want it, will be a widespread feature of tomorrow's human-centric systems and will result in a lot of software transfers among them. Think of the software as capturing your information personality and becoming nomadic, so it can roam onto whatever device you want to use.

Applications software—from word processors to Web browsers—and the way it is distributed will also change, due to economic reasons, but not the ones we have been hearing about. For years, people have been saying that the low marginal cost of copying software would drive its price to zero. This hasn't happened because software makers have been changing their products annually, mostly to keep making money through new features. This trend, and the growing ease with which nomadic software will move over the Net, will cause us to gradually stop buying the familiar shrink-wrapped software packages. Instead, we will "rent" the programs we need by having them periodically downloaded from the Net for a fee. The result of these trends is inevitable: The entire software enterprise will evolve from a product business to a service business. You'll pay a monthly fee to your software service provider, who will ensure that your software needs will be met, often automatically without you being aware of the upgrades . . . as long as you keep up with the payments. And software revenues, instead of going down, will become steady and even rise.

The ease of moving software through networks, by the way, has motivated some manufacturers to hail the arrival of so-called network computers, a new breed of inexpensive boxes largely devoid of programs and bells and whistles that are targeted to replace PCs. You will fill them with software retrieved on the fly from the Net. This is a laudable dream that appeals to organizations that like to manage their software centrally. But in practice, tomorrow's machines will be neither pure network machines that acquire their functions online, nor pure PCs stuffed with software from the factory. They will use a mix of local and distant resources through flowing, nomadic software, because that will best serve people's needs.

Rise of the Information Marketplace

Appliances, mobility, nomadic software, and the people that use these capabilities will not come together spontaneously and wondrously in the new terrain to create an era of human-centric computing. Nor is it enough to say that "convergence" of all media to digital form will achieve this goal. That's already here. What we need is a model of an underlying computer and communications infrastructure that will tie the elements together at a higher level, closer to what we want to do. Today's Web and Internet are not yet there. Stripped of cosmetic adjectives, they are basically used for voyeurism and exhibitionism. And I don't mean sex! I am talking about the millions of people and organizations showing off their wares for money, pride, or sharing, and the many millions who click away, peeking at these exhibitions. Much more than that lies ahead. The model toward which we are headed, which I have been forecasting for 20 years, is finally emerging: the Information Marketplace.

By 2010, over a billion people and their computers, along with some 100 billion appliances, will be interconnected. What will they all do? They will *buy, sell, and freely exchange information and information services.*

Make no mistake: The sharing of information and e-commerce over today's Internet is only the tip of the Information Marketplace iceberg. Take, for example, the "content" that the press and Wall Street were hyperventilating about throughout the late 1990s, in the wake of proposed megamergers like that between America Online and Time Warner. All the content you can imagine—TV, movies, theater, radio, newspapers, magazines, books—accounts for less than

5 percent of the world's industrial economy. On the other hand, a whopping 50 percent of that economy—some $10 trillion—is office work, or, as it used to be called, white-collar work. This includes buy and sell transactions, reviewing mortgage applications, processing insurance forms, dealing with medical information, filling and reviewing millions of government forms, teaching and learning, selling customer services, and a myriad of business-to-business services. That's *information work*—the processing of information by skilled humans, and secondarily by machines, and the delivery of that work where and when it is needed. This is barely happening over the Internet today, so no one talks about it. But it will be everywhere on tomorrow's Information Marketplace. Human-centric computing must make it easy for people to offer their work across space and time if the Information Marketplace is to reach its full potential.

By 2020, and by my reckoning, some $4 trillion of this information work will flow over the Information Marketplace, shaking up the distribution of labor. Just imagine what 50 million Indians could do to the English-speaking industrial world using their ability to read and write English and offer their office skills, at a distance, for about one-third of what the West pays today. Such a move would have colossal economic consequences, in the redistribution of work, internationally. It would also mark a poetic comeback for India, which may then be in a position to exert economic power on a nation like England that taught the Indians English to dominate them. As much as information work will flow from poor to rich, even more will flow from rich to rich—services that will be increasingly delivered via the Net because of speed and convenience. By the time this activity and the electronic commerce in goods level out, the "buy-and-sell" part of the Information Marketplace will grow from some $200 billion in 2000 to some $5 trillion annually, roughly one-fourth of the world industrial economy.

The "free exchange" part of the Information Marketplace will be just as important, because people have as much free time as work time, and they value what they do with it just as much. Already the lives of many people are affected through family e-mail; collaboration, playing, and dating; entertainment through listening to music and viewing images and videos; accessing information of personal interest; engaging in discussions about literature, hobbies, and social issues; publishing their views, and much more. These uses will grow; when I speak publicly, I always ask those people in the audience who use e-mail to communicate with family members to raise their hands. The ratio, largely invisible in 1995, was consistently over 90 percent in 2000. Many new activities will arise as well that we can't predict today.

Taken together, the monetary and nonmonetary activities of the Information Marketplace, driven by the onrush of faster computers and communications, computerized appliances, mobile gadgets, and portable software, will propel us toward a world overflowing with information and information-related activities. The question is, "How can we build this world so we are ensured of doing more by doing less?" rather than drowning in information overload and computer complexity. Only by throwing out last century's model for computing and adopting—indeed, demanding—a new computing philosophy, a new master plan, that lets people interact naturally, easily, and purposefully with each other and the surrounding physical world.

Human-centric computing will transform today's individual computers, the Internet, and the Web into a true Information Marketplace, where we'll buy, sell, and freely exchange information and information services using systems that will talk with us, do things for us, get the information we want, help us work with other people, and adapt to our individual needs. Indeed, it is these five basic capabilities of computer and communications systems that are the pivotal forces of human-centric computing.

As builders of computer systems start turning these forces into useful technologies, the rest of us who are collectively frustrated by today's computers can accelerate the process by tirelessly repeating the rallying cry of human-centric computing: "Information technology should help people do more by doing less!" If we shout loud enough, entrepreneurial companies will make this request their goal. They will recognize the huge, pent-up demand for human-centric systems, and will build them, upstaging the massive computer-communications establishment and shifting the market in their direction.

Much as we like to tout it, the Information Revolution is not yet here. It started innocently enough in the 1950s with a handful of laboratory curiosities dedicated to mathematical calculations. The 1960s brought time-shared computers, each used in round-robin fashion by tens of people to spread the high computer cost. Universities and other organizations soon discovered that the real benefit was not the money saved but the information shared through e-mail and document transfers within each group sharing a machine. The 1970s brought the Arpanet, which interconnected tens of time-shared machines, mostly at universities; again, this was built to spread computing costs, and again, the real benefit turned out to be expansion of the community that could share information, this time to a few thousand people. The personal computer's arrival in the 1980s made computer power affordable to millions of people who used their machines for office work and for play at home. The Ethernet, which arrived at the same time, made possible the interconnection of hundreds of PCs in local networks, mostly within organizations. The growing demand to bridge together the thousands of these local nets was addressed by the Internet, which had been already developed as a method for interconnecting networks of computers. These changes increased the community of people who could share information through e-mail and file transfers to a few million people. Then, in the 1990s, when networking advances seemed to be leveling out, and it looked like nothing big could possibly happen, the biggest change of all took place—the World Wide Web arrived as a software application for Internetted computers. It hit the steadily growing community of interconnected users with a quantitative and qualitative jolt. Creating and browsing Web sites captivated the world so much that the number of interconnected users shot up to 300 million by the end of the 20th century, as they and the rest of the world began experiencing the awesome socioeconomic potential of the Information Marketplace.

Unlike the Industrial Revolution, which has run its course, the Information Revolution is still growing. All we have today is several practical activities, an abundance of exciting promises, and a gigantic tangle of complexities, confusions, and fads—to be sure, a revolution in the making, but one that is unfinished. The missing ingredient is human-centric computing. To put it into action requires three big steps: changing the mind-set of users and designers; ensuring that our machines are easier to use and make us more productive; and insisting that new technology reach many more people.

Integrate Computers into Our Lives

The need to change the mind-set of computer users and designers sounds obvious, but we are marching in the opposite direction. Everywhere we turn we hear about almighty "cyberspace"! The hype promises that we will leave our boring lives, don goggles and body suits, and enter some metallic, three-dimensional, multimedia, terabyte-infested, gigahertz-adorned otherworld.

To which I respond with the technical term: Baloney!

When the Industrial Revolution arrived with its great innovation, the motor, we didn't leave our world to go to some remote motorspace! On the contrary, we brought the motors into our lives, as automobiles, refrigerators, drill presses, and pencil sharpeners. This absorption has been so complete that we refer to all these tools with names that declare their usage, not their "motorness." These innovations led to a major socioeconomic movement precisely because they entered and affected profoundly our everyday lives. People have not changed fundamentally in thousands of years. Technology changes constantly. It's the one that must adapt to us.

That's exactly what will happen with information technology and its gadgets under human-centric computing. The longer we continue to believe that computers will take us to a magical new world, the longer we will delay their natural fusion with our lives, the hallmark of every major movement that aspires to be called a socioeconomic revolution.

Once we change our mind-set in earnest, we will no longer put up with the maddening computer faults we now suffer. And we will be careful about what we accept from the proselytizers of technology. No longer will we be seduced by fancy buzzwords like "multimedia," "intelligent agents," "push-versus-pull technologies," "convergence," "broadband," "gigahertz" and "gigabytes," and a few hundred others already with us and yet to come. Instead, we will behave more like we do when we shop for a car: "Rather than tell me how fast the engine turns or whether it has an overhead cam, tell me about how many people it seats comfortably, the gas mileage it gets, and its annual maintenance cost." We must begin asking the same kinds of questions about computers and software: "Rather than tell me about all its gigas of processor speed and memory, tell me how quickly it can find and show me any movie I want to see, or help me find a replacement part for my lawn tractor."

As users, we want to know how much more we can achieve with a given machine or software, and at what effort, compared with what we are doing now. We'll accept quantitative or qualitative answers, as long as they address these kinds of questions. First, we'll be told that computers are different and don't admit to such measures. Nonsense. If we insist, designers and manufacturers will be compelled to respond. As they do, they will gradually adopt the new mind-set too. Eventually, they will be anxious to innovate, develop measures of usefulness, and brag about the real utility their products and services bring, versus that of their competitors.

And when the computers "vanish," as motors did earlier, we'll know the Information Revolution has finished!

Give Us a Gas Pedal and Steering Wheel

The second step toward doing more by doing less is to raise the level of controls we use to interact with our systems, from their current, low, machine level to the higher human level where we operate.

Since computing began, designers and users have been catering to what machines want. Engineers design to suit what the computer, communications system, or peripheral needs. They then throw all the components at the users and expect them to make everything function together. Miraculously, we accept without protest!

As you sit in front of your computer trying to bend it to your wishes, I imagine you trying to control a very early vintage car. Instead of having a steering wheel, brake, and gas pedal, you must wear a ring on each finger. Each ring is connected with pulling cables to levers that control spark advance, fuel mixture, the valve clearance of each cylinder, the angle of each wheel, the tension on each brake drum. What you want to do, at the human level, is go from Boston to New York. But to get there you must operate at the machine level, wiggling all the wires and levers. The prospect is so harrowing you would not be willing to undertake the trip. Yet we do it every day when we fire up the computer. We need to replace the low-level controls with the equivalents of the steering wheel, gas pedal, and brake.

Finally, computers will be easier to use and make us more productive if we can stick to a few common and consistent commands to do what we want with information, regardless of where the information resides. It's inconceivable to me that we are still using different commands between operating systems and browsers, just because operating systems work on information that is local to our personal computers while browsers work on distant information that sits on the Web. In both cases we want to do exactly the same things: enter information, see it or hear it, move it around, transform it, use it as a program to accomplish a task, and so on. Human-centric computing requires that we have the same set of commands for both of these cases, as well as for other gadgets and auxiliary systems that, inevitably, do the same things with information. This situation is as ridiculous as using your steering wheel to turn your car on city streets, but having to use the brake pedal to turn the car out in the country. Today's systems not only force us to learn different commands, but also entirely different ways of working each time a system changes or is "upgraded."

Many people confuse wishes with claims. Computer vendors have abused the phrases "ease of use" and "user-friendly." What they usually mean is that you can change a few colors or icons on the screen, which is supposed to give the impression that the system is bending to your commands. Such feeble cosmetics are tantamount to painting a smelly trash can in pretty colors to chase away the bad smell. You would be better off if all the multicolor, multimedia bells and whistles were replaced by a thin, noisy pipe, through which you could speak with a wise old man at the other end. Unfortunately, we do not know how to make machines behave intelligently, except in extremely limited contexts. Nor can we create "intelligent agents"—another darling of the spin doctors—that can act in our stead, behaving the way people expect an intelligent surrogate to act. When I say we must improve ease of use and increase productivity, I mean improve the fundamental communication between people and machines, not wax commercial about unrealistic desires.

We have complicated things enough. It's time we change our machine-oriented mind-set and invent controls that are much closer to what people want to do. We need the steering wheel, gas pedal, and brake of the Information Age. New technology can help us in this quest. And that's a good part of what this book is about.

Reach All People

The third step needed to make computers human-centered and help us finish the Information Revolution is to reach more people. Many more.

At the beginning of the 21st century there were some 300 million people interconnected over the Internet. That big number makes us feel pretty smug. Yet it represents only 5 percent of the world's population. It's scandalous to characterize the Web as "worldwide" when it spans such a tiny portion of humankind. The voices of billions of people in the developing world and the poor regions of the industrial world cannot be heard through anything other than television news tidbits and government information feeds.

If we do nothing, matters will get worse. The rich, who can afford to buy the new technologies, will use them to become increasingly more productive and therefore even richer. The poor will be left standing still. The outcome is inescapable: Left to its own devices, the Information Revolution will increase the gap between rich and poor nations, and between rich and poor people within nations.

This gap is already huge. In the U.S. economy, an average of $3,000 in hardware, software, and related services is spent each year per citizen. In Bangladesh it's $1, according to that country's embassy. I suspect that if I could find an "embassy" representing poor Americans, or the poor of any industrial nation, I would get an equally screeching dissonance between information technology expenditures in the ghetto and the suburbs.

Some people believe the gap will close by itself, because of the growing reach and potential benefits of the Internet. It can't. The poor could have a crack at these benefits if, somehow, they were provided with the communications systems, hardware, software, training, and other help they need to join the club. Absent such help, they can't even get started.

We cannot let this gap widen. It's high time we begin closing it. Not just to be compassionate, but also to avoid the bloodshed that, historically, follows every widening rich-poor gap.

This may sound like a worthwhile social goal, but not something that will necessarily help the rest of us. Not so. First of all, if engineers begin to design computers so simple that they can be used easily by people with limited skills, the machines will be easier to use for everyone! The World Wide Web Consortium is already using this important principle in its Web Accessibility Initiative, which is creating technology to help people with visual, auditory, and other impairments to use the Web. These improvements also make the Web easier to use for people without these limitations. The history of technology shows many more examples like this; whenever designers build utility for the least-skilled user, they enhance utility for all users.

Second, if we can increase the number of people who will benefit from the technologies of information, the productivity of the entire planet will rise. New technologies will not only help the poor become literate, learn how to plant, and take care of their health, but will also help them sell their goods and services over an expanding Information Marketplace. The potential is immense. Companies in developed economies could buy information work from people in less-developed countries at greatly reduced prices, as is now done with manufacturing. Entrepreneurs in developing countries could even help those in developed countries. Imagine a new breed of useful counseling exchanges between the rich people of the West, who are often troubled by depression, divorce, and family problems, and the poor people of the East, who seem to counterbalance lack of money with strong family ties and inner peace: Older, experienced Indian women could spend a lot of time over the Net chatting with Western divorcées, who could benefit from their advice at costs substantially below the psychologist's counseling fee. The lack of time that characterizes Westerners would be counterbalanced by the plentiful time of people in India.

There is little we can do in this book to help people become interconnected, except call attention to the disparity. Yet we must not let this important objective be forgotten, for it is essential to our broader quest. We must also persist because the Information Marketplace is huge and largely unexplored. If even a small number of Nepalese or a few inner city people found a way to become productively interconnected, they would serve as role models to their peers. A timid experiment could turn into a beneficial economic spiral.

Some people have an overarching fear about what computers may do to us. They believe that increased deployment of accessible machines will merely accelerate our becoming robotlike freaks who are driven by efficiency, instead of by the timeless pursuits and relationships that make us human. Other people are convinced that better use of information technology will free us from what is already an "inhuman" way of living and let us focus on what's truly important to us. Is it possible for human-centric computing to enhance our humanity? Or does the horizon for "doing more by doing less" end at greater productivity?

On to the master plan.

TWO
LET'S TALK
NATURAL INTERACTION

Our computer systems are hard to use. They enslave us rather than serve us. If we do nothing things will get worse, as billions of people and physical devices become interconnected. We need a radical change to a new breed of human-centered computers. We must simplify our computers in a big way.

That's a great wish. But what do we really mean when we say we should "simplify"?

Our instinctual reaction is to equate simplicity with leanness—of features and of controls. That's certainly a good avenue to explore. The Web has already shown the power of this approach; its single control—a mouse click on any highlighted phrase or image—has captured with its simplicity and ease of use hundreds of millions of enthusiasts. We should throw away 90 percent of the features and controls that come in today's bloated software.

But cutting down the number of controls isn't the full story, and can even lead to problems. For example, a typical digital watch has two buttons. One changes modes. Press it once to set the time, again to set the date, again to set the alarm, another time to set the day of the week. The other button lets you scroll to a specific time, date, or other information you want to set, whatever mode you may be using.

Even though the watch has only two controls, this system is not simple, because it causes people to become confused and forget which mode they are in, and which procedure they should follow.

Maybe it's not the number of controls that should be reduced, but the many different functions a system can perform. That sounds promising until you imagine a car that can only carry out two movements—go forward or not, and turn right or not. In principle, you can drive this car anywhere; if you want to turn left, you just keep turning right until you point to the left. But that ridiculous contraption is not simpler to use than your current car, which has many more capabilities, like accelerate, brake, and so on. So minimizing the number of capabilities a system has isn't really what we mean when we ask for simplification.

How about configuring a system with a control for every conceivable action we might want to take? That won't do, either. Such a system would confuse every one of us, and result in unwieldy manuals and unwanted interactions among endless features.

Specialization can help simplification. The bottle opener and shoehorn symbolize this time-honored tradition. In the Industrial Era we gave custom-made, motorized tools to doctors, mechanics, plumbers, and gardeners that simplified their jobs. But in the Information Age those same groups of people, plus musicians, accountants, engineers, and lawyers, are trudging along with the same word processor, spreadsheet, database, and graphics editor. Specialization has promise in the world of human-centered computers, but we are searching here for overarching ways to simplify computer systems, before specializing them through applications.

Maybe we are after an impossible task. I often hear the excuse that software systems are inherently so complex that they cannot be simplified. Nonsense! This is the mind-set of lazy engineers. Since time immemorial humans have strived to simplify the world so they can understand it. At a time when we are beginning to unravel incredibly complex natural systems like our genetic makeup, it's outrageous to claim that the software we put together is too complicated to simplify.

Perhaps our call for simplification means we want a system whose machine actions match our human intent. When we use a VCR, our intent is to tape a specific show. Instead, we struggle to set the start and stop times of the recording, the recording speed, the channel number. Programming the machine in this way is so far from our level of interest and so frustrating that the experience has become the butt of endless jokes. We should be able to command "Tape tonight's ball game" and be done with it.

Saying that a system should let us carry out our intent at our level and with little effort is tantamount to saying that it should be natural to use. You were not born with a keyboard or mouse socket in your belly button to communicate with the world around you. You were born with ears, a mouth, eyes, and the ability to listen, speak, and see. Human-centric computing begs that you use these natural human capabilities to communicate with your machines so you won't have to burden yourself learning new, complicated approaches to do what you already know how to do effortlessly.

Finally we have hit upon the biggest lever of system simplification: natural interaction with machines. It is the first step toward human-centric computing.

Here is an example of the kind of natural interaction we can expect by 2005. It is an exact replica of a spoken dialogue between human and machine, using a Year 2000 advanced prototype called "Mercury" at the MIT Lab for Computer Science.

It's late Tuesday night and duty calls. You have to make a sudden business trip. You need to book plane flights right now. So you call the automated Mercury airline reservation system. It picks up, and a machine talks to you.

What is your name please?

Victor Zue.

Welcome Victor Zue. You have successfully logged onto Mercury. How may I help you?

I want to go from Boston to London on November 7.

Okay, from Boston to London on November 7. Can
you provide the approximate departure time or
airline preference?

I'd like to fly United.

Sorry, I'm not sure what you said.

I want to fly United.

Okay. United flights from Boston to London on
November 7. One moment please . . . I have a
nonstop United flight leaving at 7:30 P.M. and
arriving at 7:10 A.M. the following day.

Please book this flight.

United Flight 998 has been added to your itinerary. You depart Boston at 7:30 P.M. and
arrive in London Heathrow at 7:10 A.M. the
following day. Where would you like to
go next?

I want to go to New York City two days later.

Okay. United flights from London Heathrow
to New York City on November 10. One moment
please . . . Can you provide a departure
or arrival time? The earliest nonstop
leaves at 9:00 A.M. and arrives at
12:15 P.M.

Please book this flight.

United Flight 957 has been added to your itinerary. You depart London Heathrow at 9:00 A.M. and arrive in New York's Kennedy at 12:15 P.M. Where would you like to go next?

I want to return to Boston that same evening.

Okay. United flights from New York's Kennedy to Boston departing between 5:00 P.M. and midnight on November 10. One moment please . . . I have three nonstop flights, a flight departing at 5:30 P.M., a flight departing at 7:25 P.M., and a flight departing at 9:10 P.M. Please select one.

Please book the second one.

United Flight 7282 has been added to your itinerary. You depart New York's Kennedy at 7:25 P.M. and arrive in Boston at 8:47 P.M. Shall I price your itinerary?

Yes, please.

Pricing your itinerary will take a while. In the meantime, I'll summarize your itinerary for you. On November 7, you depart on United Flight 998 from Boston at 7:30 P.M. and arrive in London Heathrow at 7:10 A.M. You arrive one day later. From there, on November 10, you depart on United Flight 957 at 9:00 A.M. and arrive in New York's Kennedy at 12:15 P.M. From there, you depart on United Flight 7282 at 7:25 P.M. and arrive in Boston at 8:47 P.M. The price for your itinerary is $1,404.95. Shall I e-mail your itinerary to you?

Yes, please.

> Your itinerary has been e-mailed to zue@xxx.
> Shall I start a new itinerary for you?

No, thank you.

> Thank you for calling. Have a nice day.

Elusive Intelligence

Upon hearing this kind of scenario about a computer system that carries out our commands, any journalist or lay person immersed in the Internet will jump up and say, "What you really need is an intelligent agent—an electronic servant that will do what you want in your stead." That's absolutely true. Such a piece of software would be as natural as having another human being next to us, and it would represent the greatest simplification possible. Give it to me and I'll throw away every computer system I own and every new system I am designing. The problem is that despite the incessant reference to "intelligent agents," as if you could go to the corner drugstore and buy them, no one has built them and no one knows how to build them.

A field of research called artificial intelligence was established in 1956 by scientists from Carnegie-Mellon University, MIT, and Stanford University with the twin goal of making machines behave intelligently and understanding how people think. The field is still going strong and has resulted in several innovations now considered in the mainstream of information technology. But the first goal—the injection of humanlike intelligence into machines, known as "the AI problem"—has eluded solution by some of the world's best scientists and technologists for nearly half a century. And no such solution is discernible on the horizon. No one has been able to imitate by machine the common sense exhibited even by the average toddler.

The "intelligent agents" touted at the turn of the 21st century have been mostly programs that carry out a thin sliver of elementary, humanlike logic via what computer scientists call if-then-else procedures. For example:

If the car phone rings and if the radio is on, then mute the radio.
If a call is initiated and the radio is on, then mute the radio.
If the phone is hung up and the radio is on, then unmute the radio.

If a program like this shows a tiny portion of humanlike behavior, it is dubbed "intelligent," usually for marketing purposes, reinforcing the illusion that intelligent agents are commonplace. But even the most advanced programs constructed to date in various labs can behave in a marginally humanlike way only in a very narrow context—like the Mercury system did for airline reservations. That is very useful for human-centric computing. But it falls far short of the breadth of understanding insinuated by the ambitious term "intelligent agent." Let's not fall prey to the syndrome of accepting a wish, stated with a fancy name, as an established capability.

The future prospects for machine intelligence are unknown, as are the fruits of all high-risk, high-payoff research. There are a lot of philosophies, approaches, and beliefs, but no one can responsibly state how far we'll be able to go toward emulating by machine the intelligent behavior we normally associate with people. That does not diminish the importance of looking for answers. The problem is central and merits more attention than it is getting today, as a result of past disappointments. The kingpin of machine intelligence is machine learning—the ability of a machine to learn from its "experiences," as it goes along, rather than relying on a human programmer to tell it how to behave intelligently.

With no assurance that machine learning and machine intelligence will happen, we must set aside such wishful thinking and move along with human-centric approaches that will help us interact with our machines naturally—with speech and vision.

Speech and Vision: Different Roles

Speech and vision are the two principal ways we have used to interact with other people and the world around us for thousands of years. And since vision occupies so much more of the human brain than speech, we may be tempted to declare it the queen of human machine communication. That would be an easy—but deceptive—conclusion. Vision and speech do not serve the same natural roles in human communication.

Being Greek, I can still hold a "conversation" in Athens through a car window, using only gestures and grimaces—one clockwise rotation of the wrist means "how are you," while an oscillating motion of the right hand around the index finger with palm extended and sides of mouth drawn downward means "so, so." A sign language like American Sign Language works even better. But when speech is possible, it invariably takes over as the preferred mode.

If we take a closer look, we see a puzzling asymmetry: We use speech equally for two-way communication. But vision is used mostly one-way—for taking in information—and only secondarily for generating visual cues that reinforce spoken communication. (Visual communication would have been a two-way proposition, too, if we were born with built-in display monitors on our chests.) Why this difference? Perhaps the one-way power of vision was nature or God's way of ensuring survival in a world of friends and enemies, edible and man-eating animals, useful and useless objects, lush valleys and dangerous ravines, where maximum "information in" was essential.

But then, why didn't nature or God make speech just as powerfully a one-way capability as vision? I'll venture that speaking and listening were meant for a different purpose—intercommunication, where, unlike survival, a two-way capability was essential. And since survival was more important than chatting, the lion's share of the human brain was dedicated to seeing.

These conclusions run against the common wisdom, especially among technologists, that for human-machine communication, "vision is just like speech, only more powerful." Not so. These two serve different roles in our natural selves, which we should imitate in human-centric computing. Spoken dialogue should be the primary approach for exchanges between people and machines, and vision should be the primary approach for human perception of information from the machine.

We can imagine situations where a visual human-machine dialogue would be preferable; for example, in learning by machine to ski or juggle. But in human-centered systems, we are interested in human-machine intercommunication across the full gamut of human interests, where, as telephony has demonstrated, speech-only exchanges go a long way. (Might these basic differences between speech and vision have contributed to the lack of success of various "video-phones"?)

If we can combine speech and vision in communicating with our machines, as we do in our interactions with other people, we'll be even better off. Such a blending is beginning to happen in the research laboratories. But it's not easy to do, since the technologies for speech and vision are in different stages of development. Nor is the obvious and natural wish to combine them in human-centered computers reason enough to ignore their different roles.

Let's Talk

Speech systems have been promised for a long time, but they are finally ready to burst upon the scene. They fall into two broad categories: speech recognition and speech understanding.

In speech recognition systems a sentence, spoken by a person, is converted by the computer into text, most often in a word processor. Several such systems are commercially available today, led by L & H (Lernout and Hauspie), Dragon (now owned by L & H), IBM, and Philips. They've even reached the average user in the form of speech dictation programs, like IBM's ViaVoice and Dragon's Naturally Speaking, for PCs. People who have physical problems typing often use such systems in their work.

Typically, speech recognition programs involve a period of learning. The user trains the machine to get accustomed to his speech patterns as he repeats several sentences for about an hour. At a subliminal level, some training goes on in the other direction as well, as the human gradually learns how to adjust his speech to get better results from the machine. The recognition accuracy varies, but it can be in the high 90 percent range, according to the manufacturers' claims. I have not been able to achieve such accuracy, but even if possible, it would still result in one wrong word in every two or three sentences. The user would then have to correct the mistakes.

Speech understanding programs, in comparison, respond to a spoken sentence with action. Sometimes the action is an answer to a query like "Weather forecast for Boston, please." Other times the action may be control of a physical device, like a thermostat, in response to the spoken command "Please make it a bit warmer in here." Unlike recognition systems that are used for continuous transcription, speech understanding systems involve dialogue: You say something, and the machine responds, requests clarification, or carries out your command.

Speech understanding systems require no training by the speaker and can therefore be used by anyone. However, when used consistently by one person, they gradually adapt to that speaker's idiosyncrasies, resulting in even better performance. They, too, exhibit an accuracy in the high 90 percent range. And they can understand what you tell them and carry out useful actions, as long as you keep the discussion confined to a narrow context, like making flight reservations, navigating in a city, getting stock prices, or asking about the weather. Since the human-machine dialogue involves short exchanges, errors don't accumulate over long dictation stretches and chances are the machine gets your phrase right. When it does not, the system alerts you to repeat or rephrase the query. Or it does the wrong action, which causes you to tell it what it should do. You have reassuring feedback of the progress of your interaction, similar to the nods and occasional blank stares you receive from a human companion.

Most emerging commercial systems are of the speech recognition variety, focused on dictation. I find this strange, at a time when most office workers no longer use dictation and dictation machines are no longer in vogue. Perhaps the rationale is that people use word processing programs, and dictation is a straightforward substitute for typing. Sooner or later, however, this focus will shift toward speech understanding systems, because they are the ones that meet the objectives of human-centric computing—we interact with them via spoken dialogue, as we do with the people around us. We don't dictate messages to each other!

The Mercury flight reservation system, developed by Dr. Victor Zue and his team at the MIT Lab for Computer Science, is a prime example of a dialogue-oriented exchange. Here is another, which also can be operated from a distance by phone.

```
Hi. Welcome to Jupiter, the MIT Lab for Com-
puter Science weather information system. This
call is being recorded for system development.
You may hang up or ask for help at any time.
How can I help you?
```

Weather forecast for Boston, please.

```
In Boston, Monday, mostly sunny, high in mid-
60s. Monday night mostly clear, lows 45 to 50.
What else?
```

How about Athens, Greece?

```
In Athens Monday, sunny skies expected. High 68
and low 57. May I help you with something else?
```

Yeah. Give me that in Celsius.

```
In Athens Monday, sunny skies expected. High 20
Celsius and low 14 Celsius. What else?
```

What places do you know in England?

```
I know of the following places in England: Cam-
bridge, Liverpool, London, London slash
Heathrow, and Manchester. Please select one.
```

No. Good-bye, Jupiter.

```
Thank you for calling. Have a nice day.
```

The Jupiter weather system goes beyond converting into text what the user says. It understands what the user wants to know about the weather, and responds accordingly. Besides giving forecasts, weather advisories, and data such as temperature or sunrise, Jupiter has an understanding of time, like "today" or "long range," and of space, like "all the places in England." Jupiter also has a sense of its state—if you ask for the weather tomorrow in Paris and then ask for the sunrise there, it will assume that you are still interested in tomorrow. Jupiter continuously gets weather information from the Web and from satellite feeds, so it is always current to within minutes of the latest posted forecasts.

The Mercury system for booking airline flights has several additional capabilities. It asks for the user's name, so it can put in play that person's preregistered profile, which lists preferences such as class of service, seating, and other custom choices. Asking for the user's name also helps the system certify that he or she is authorized to use the system and has a valid payment arrangement. Mercury involves a longer dialogue than Jupiter because it is building a plan that depends on several travel legs. This can make it difficult for the user to remember all the information that he has decided upon, so Mercury ships him an instant e-mail of the itinerary, price, fare codes, and other information. Many travelers at our lab use Mercury to plan their trips. For now, they forward the plan to a human travel agent for final processing and billing; but were the system commercial, all this would be done automatically.

The same LCS group is building several other speech understanding systems. *Pegasus*, now under construction, reports the status of airline flights, including delays, arrival times and gates, and changes in these up to the last minutes of a plane's taxiing at the airport. *Voyager* helps people navigate in a city. It provides up-to-date traffic information about the main thoroughfares, and locates restaurants, museums, banks, and other landmarks on a map that is accessible via the Web. *Orion* is a more sophisticated system that takes brief spoken instructions, like the date and destination of a journey, then carries out all the associated actions we would do, on its own. Orion goes beyond being a query-response system to one that increases human productivity through automation.

I've provided these examples from LCS because I know them best. Many new start-ups, like Nuance, SpeechWorks, ViVo, iPhrase, and NetBiTel, are joining the big-company quartet of Dragon, IBM, L & H, and Philips in the quest to get an early share of what promises to be an explosive market for spoken language systems. Several varieties produced by these companies will appear by 2003. The hardware is in place, since even the microprocessors used in run-of-the-mill PCs are fast enough to process speech. The tougher nut to crack will be creating new systems tailored to new applications. Large and complex, these software systems cannot yet be built rapidly and routinely. They require 5 to 10 person-years of work, and a good deal of individual attention, before they can reach the level of maturity shown in Mercury or Jupiter.

To overcome this difficulty and enable the creation of speech understanding systems in new areas, Zue's group has started building "light" systems—portable programs that work in even more narrow contexts and can be put together rapidly by people with ordinary skills. They might let you control all the functions of your home stereo or car radio—turning it on, tuning in a station, putting that station into memory, and adjusting the volume. All so you can wander around the room . . . or keep your eyes on the road. You could customize commands, such as "kill it" for "off," to which you could later add "shut up," or whatever other verbs suit your fancy. Such lightweight systems will make their commercial appearance even sooner. Late in 1999, for example, a few speech-driven radio-tape-CD players had already appeared in upscale electronics catalogs. Lightweight systems will be particularly helpful in the automation of human tasks, and in accessing information.

Let me complete this topic with a brief comment about speech synthesis—the ability of machines to produce speech. This is easier than speech understanding, but for years speech synthesizers have had a tinny, nasal, otherworldly tone that has become the caricature of "machine speak" in science fiction movies. I regret to inform sci-fi lovers that the newest speech synthesizers sound wonderfully natural. This is achieved by gluing together prerecorded speech fragments produced by a real person, in ways that makes the "joints" imperceptible. Complex issues like breathing, accenting, and turning assertions into questions are handled well. The results are so impressive that new synthesizers designed for restricted contexts are hard to distinguish from a real person. These developments will help our human-machine exchanges feel even more natural.

Companies and organizations that want to harness the productivity benefits of human-centric computing through natural interaction can get a head start by planning where they would use speech systems. They could then test nascent systems as they begin appearing. By doing so, they will be ready to leverage the technology fastest when it becomes fully commercial. Taking this tack will help them shake down the technical, business, organizational, and logistical aspects of using speech in the enterprise.

The best prospects are activities that provide high utility and are light in speech requirements. The context of discussion should be very narrow with a relatively limited vocabulary, like the weather and airline systems. Consider how many different "bins" or "buckets" the incoming queries may fall into, regardless of how they are phrased. For example, there might be three initial buckets in a speech system that lets customers interact with a product catalog: find more information about a product, buy it, or return it. A good system would not start by saying, "Do you want to get information on a product, buy a product, or return a product?" It would listen to the customer's opening line and try to recognize which of the three bins it belongs to. If it understands a buy request, it would acknowledge by repeating, and advance with a query:

```
I understand you want to buy a product. What
kind of product are you interested in?
```

At that point, the user might say, "No, I don't want to buy anything," or "I am interested in sweaters," thereby directing the dialogue in the desired direction. The speech system would then continue the dialogue, by product or category or both. If approached in this way, the number of buckets can grow to the tens or hundreds, at any given level of the discussion, which is a reasonable context for systems now under construction. If the buckets end up in the thousands, though, the application may be beyond current capabilities.

Similar systems could be tried for completing a sale, locating a package in transit, quoting a stock price, or providing information about the business or organization. Speech systems might also be used internally to report activities of subsidiaries, or to update a master location database of employees traveling on business. Governments could give callers information about tax forms, the status of a refund check, or regulations. Let's hope that tomorrow's speech understanding systems will spell the end of those murderous, $100 automatic answering systems that force us to listen to endless lists, push buttons, and get nowhere!

The arrival of commercial speech systems will not cause keyboards to vanish. They have their proper role in entering text and numbers, or where accuracy or quiet are at a premium. The same holds for handwriting on a portable's screen. These older devices will persist but be reduced in number.

The impact of speech systems will extend well beyond their role as input-output devices for the industrial world. They will help satisfy a major human-centric objective—increasing the number of people in the world who can use the Information Marketplace. For example, the Chinese could use this technology to speak to their machines, without having to resort to ideograms. Although keyboards for typing Chinese ideograms are far more complex than those for typing English, experimental speech understanding systems at the MIT Lab for Computer Science for Mandarin Chinese are no more complex than those that understand English. Unlike typing, speech understanding by machine is equally practicable for both languages.

Speech technology could also help people anywhere in the world who cannot read or write, but who could still have productive exchanges on the Internet using their native speech.

More must be done, however, to enable non-English speakers to interact with today's predominantly English-language Internet. The desire to do so is huge, as is the accompanying frustration. I'll never forget the undercurrent I encountered at a March 2000 Taipei conference entitled "Creating a Chinese Language Based Internet Economy." The barely hidden feeling among the people there was: There are one billion of us. We should build our own Internet and lock out the English language Internet!

That won't work, of course, if for no other reason than the Chinese sell so many manufactured goods to the West. But the Information Marketplace must be a truly international medium that is equally useful to all its participants. The most promising avenue to internationalization is the ancient human practice of translation, but with an important twist that I call "total translation." By this I mean not only a conversion of a Web site's sentences from one language to another, but also a "translation," to the extent that it's possible, of the culture and mind-set inherent in the site to the culture and mind-set of its new audience—a difficult and imperfect, yet essential, task.

Here's how this approach would work. People with superior knowledge of at least two languages, and the associated cultures, would form a new breed of dot.coms that would offer total translation services to organizations in each of their linguistic territories. A Chinese company, specializing in Chinese and English, would sell its services to Western companies anxious to do business in China, and to Chinese organizations seeking visibility in English-speaking countries. The translator companies would thrive, because the economic motives toward universal visibility and reach are powerful. So much so that they could overflow beyond the commercial sector to help the spread of noncommercial multilingual sites.

With well-translated sites, and speech understanding systems in place, tomorrow's Information Marketplace could open its doors to perhaps a billion more people, who would otherwise be locked out. That would be a dramatic gain in the human-centric quest to increase the number of interconnected people worldwide.

Show Me

To ensure natural interaction, human-centered systems should have powerful displays that can convey a maximum amount of information to us. And as machine vision progresses, systems should be augmented with the ability to see the gestures and grimaces we make that reinforce our spoken words.

The visual display of information is in good shape and will get even better. Going beyond the traditional, two-dimensional monitors, projection screens, and flat panels will be difficult, but worth the work. Ten-foot, and even bigger, displays are in the works, made by stitching together several projected images, and actively controlling the projectors so that the stitches are invisible. Three-dimensional displays have been built in the lab, but they are costly, their fidelity is not great, and they typically require the viewer to wear special glasses. Holographic and moving mirror techniques can eliminate the glasses, but are even further away from commercialization. An approach using millions of miniature lenses glued on the surface of a large, flat display looks promising for 3-D display without glasses. Special chambers—"caves" where images are projected all around you—offer powerful visual experiences but require substantial investment. Synthetic cameras offer a new kind of magic: Images from a hundred stationary cameras around a football field are cleverly stitched in your computer, to provide a display of the game from whatever viewpoint you choose—your favorite player's forehead, or the ball itself.

Head-mounted displays—goggles and helmets that project engulfing computer-generated images and also feed back to the computer information about your head and eye positions—are used for virtual-reality (VR) experiences like games, designing an airplane cockpit, and visiting a potential vacation site. The images are still jerky because of the huge amount of computation required to calculate how the scene should change in response to movements of the wearer's head, a deficiency that often causes motion sickness.

A variant of VR, called augmented-reality (AR) displays, may prove more useful in helping us do more by doing less. The AR glasses let you see your surroundings naturally. But then a computer-generated image can be superimposed on them. A plumber might call up a schematic of the water pipes inside a bathroom wall, and look through this display as he looks at the actual wall, so he can "see inside" the wall and know where to saw a hole to fix a leak. A surgeon about to excise a cancerous tumor deep inside a patient's brain is guided by an image of the tumor, generated by an MRI scan, superimposed on the actual patient's skull she is cutting. Over 400 such operations have already taken place at Boston's Brigham and Women's Hospital using a system developed by Eric Grimson of MIT, in collaboration with the Surgical Planning Laboratory at the Brigham. The University of North Carolina, NASA's National Bio-computation Center at Stanford University, and the Center for Computer Integrated Surgical Systems and Technology, at Johns Hopkins University are among the many institutions developing such "X-ray vision" surgical displays.

Research projects and ideas for other display approaches abound. For example, imagine asking aloud, "Where are my keys?" and hearing the computer in your house walls say, "Under there," while its laser beam shines on your folded newspaper. Enough said! The visual communication from machines to people is already great for human-centric computing and getting better.

However, visual communication in the other direction—from you and the surrounding world to your machine—is not yet there. You can't point your camera out the window and have it tell you that your wife and children are walking toward your house but without the dog, and that your young son is proudly carrying the little red car he wanted so badly. As in the case of speech, vision systems are successfully used in very narrow contexts, like recognizing the shapes of bolts in a bin, or checking whether a known circuit pattern has been laid out correctly. But even in narrow contexts we cannot yet get the level of human utility we get from a speech understanding system. The difficulty is that vision systems must deal with much more information, and they must recognize and understand patterns in two dimensions, whereas speech systems need to extract meaningful patterns from only a one-dimensional sequence of acoustic morsels. Besides, why should vision be easier for the machine than for you, when we know it occupies a much bigger part of your brain than speech?

There is interesting progress despite the difficulties. Paul Viola of the MIT AI Lab has built a system that can recognize the face of a person (out of 15 or so people) approaching the lab's door. If the face belongs to a certified lab member, the door opens. If not, the guest is directed to register her affiliation, purpose—and face. Other face recognition systems have been demonstrated by Alex Pentland at the MIT MediaLab and by companies like Visionics. Advanced humanoid robots at MIT's AI lab, and at Waseda University and the Science University in Tokyo, can recognize certain human facial expressions and respond with convincing expressions of their own. They see a happy face, they smile back; a frown, they demur.

Another way to use vision is in "intelligent rooms." In work led by Rod Brooks of the MIT AI Lab, these rooms are outfitted with tracking cameras that follow people as they move around and directional microphones that can pick up what they say, regardless of where they are. A person could call for a map to be displayed on a large monitor, point with his finger to a place on the map, ask that the monitor zoom in on it, then ask for display information about it. Coarse gaze detection could help, too, and is relatively easy for a machine, because the balance of white that shows on either side of your pupils tells a video camera where you are looking. Such visual interactions might make it possible for our machines to read lips, take in gestures, and watch body language, which together with speech will let them understand us better. These are important elements of human-centric computing, because they make our machines easier to use. They also break the age-old pattern of people going to their computer to get something done. You simply go about doing your work, and the computer is there for you, when and where you need it.

A different vision technique—pattern classification—may help us in other ways. For example, a computer and video cameras recording the hustle and bustle of a busy city street can deduce the traffic pattern of people and cars. Before you scream "Big Brother!" keep in mind that the cameras need not (and should not) recognize individuals, just the movements. Such a classifier, installed at your house, could help spot burglars. With a little help from you simulating a prowler, it could learn to recognize the infrared pattern of a human moving in the backyard. After dark, it could alert you when an intruder approaches—and not go off if the interloper is a wayward dog or raccoon. A similar monitor in your aging mother's apartment could alert you by phone if it detects the pattern of her falling down.

I don't expect a great impact from other natural, sensory forms of communication, except for some interesting games and novel specialties. In the field of "haptics," which deals with touch and physical motion, people grip handles or wear sensor-laden gloves that convey to a computer what they touch and press, as they feel real resistance to their actions. This research will lead to haptic interfaces useful in virtual reality and medicine, but is still a bit in the future in improving the general public's easy use of machines. Smell and taste are not yet at a level where they can help much.

Speech and vision will remain the queen technologies of human-centric interaction among people and machines.

A New Metaphor

A good way to make human-machine interaction more natural would be to develop a better metaphor. A computer metaphor is a familiar object or activity that your computer imitates with its commands, display arrangements, and behavior. The two main metaphors we have today are the desktop and the browser. In the desktop metaphor, the display screen mimics a typical desk; information is kept inside folders, which can be opened, closed, and slipped into other folders. With Web browsing, the metaphor is downtown window shopping; you gaze at various "storefronts," see one you like, and (click) you enter. Inside, there are more options to browse, you choose another, and again you enter. Like a linguistic metaphor, the power of a good computer metaphor is that it makes a new system you don't know behave like an old "system" with which you are familiar. This lets you use the new system and get useful results out of it easily, since you don't have to struggle learning new concepts and commands.

The desktop and browser metaphors are powerful, but glued together as they are in today's machines, they make no sense because they work in different ways. Their makers claim they have to be different, since the operating system controls a computer while a browser controls a communications network. But that kind of excuse didn't stop the telephone from looking and feeling the same to us, regardless of whether we make a local or a long-distance call, even though different equipment is involved in local and long-distance telephony.

We wouldn't tolerate a telephone system that forced us to use one dialing procedure and keypad for local calls but a different procedure and keypad for long-distance calls. Yet we do with computers. We are forced to use the operating system's desktop for "local" information on our own computer, and the browser for distant information on the Internet, even though we want to do the same things with information in both cases. Why confuse us with two different metaphors?

Human-centric computing demands that we merge the metaphors into a single system. A few developers are making cosmetic changes by adding a couple of features of one to the other. But no one is going beneath the surface to create a unified, single system. One reason for this is competition—rivals would have to cooperate or abandon a lucrative operating system or browser. Another reason is politics: The Justice Department's two-year prosecution of Microsoft was built in part on charges that Microsoft was combining its Explorer browser with its Windows operating system, which posed unfair competition for independent browser makers. It's amazing that while we heard a great deal about every conceivable rivaling corporate interest, the far bigger human interest of ease of use was ignored. Even the judge's Finding of Fact assumed that the division into browsers and operating systems was cast into concrete, ignoring the powerful "fact" that the boundaries between software systems should change to improve human utility.

But even a perfect system built from the ground up, one that captures what browsers and operating systems do today, wouldn't be adequate. As we'll see, there are many more things that our machines will do. And technology is advancing all the time, bringing new capabilities to the fore. Consider the millions of interconnected appliances that we will want to control. How will we "grab" them and "tell" them what to do? What's going to be the metaphor? Giving them an address on the Web, as some manufacturers are beginning to do, and saying "Go get them," is not enough. That's like saying: "Everything in this world is in some physical location. Go get it." Organizing these appliances according to what they can do will be more helpful.

The challenge for tomorrow's systems builders is this: Find a new metaphor that captures not only what people can do with local and distant information, but also the new human-centric capabilities we want from machines in tomorrow's computing terrain. Such a metaphor would go a long way toward making tomorrow's systems easier to use.

One favorite and much-discussed metaphor is the assistant or servant. Think of your computer as an obedient servant that can understand your wishes and is familiar with your habits. You speak to it in your native tongue and it dutifully carries out your commands. Compelling, but a pipe dream. This puts us squarely in the domain of intelligent agents, which we do not know how to construct.

How about a virtual geographic metaphor? I favor this one because I dreamed it up . . . no doubt along with many other people. Information sites are organized as floors of various buildings, which sit along streets and avenues, aggregated into towns and cities, all shown on navigable maps on your screen. You zoom into and out of buildings, down alleys, across towns. The geographic mapping could be realistic; if you want to go to the Louvre, you navigate to Paris, and then to the museum. A more exciting prospect is to arrange your own information in a "virtual" map of your own creation. There could be a street of stores you frequent, or a town known for its off-color sites. Shopping avenues might run north and south according to the category of goods sold—Foods Avenue, Clothes Avenue, Household Goods Avenue, Electronics Avenue, Music Avenue, and Books Avenue.

An interesting variation of this metaphor adds the notion of time. Moving your joystick forward, back, right, and left propels you north, south, east, and west, but pushing the joystick down takes you to the past of whatever site you are visiting, and lifting it up moves you to its future plans. Such a metaphor, which I have dubbed a "historicopter," would be fantastic for studying the world, with every country contributing its history, current events, and future plans to the experience via its Web sites.

To be complete, these geographic metaphors would have to be augmented with the actions that you could carry out on information, once you found it—for example, watch it, hear it, print it, change it, or run it as a program. Successful virtual information maps might be sold, traded, or given freely to people by organizations. They are technically straightforward to implement and would be easy to master, even for novice users, because moving in physical space, like talking and seeing, is a natural human experience with thousands of years behind it—a great asset for human-centric computing.

Other metaphors may be even better. Some people say we should have several metaphors, one for each occasion we use our machines. That's technically pleasing, but not as economical and easy to remember as a single metaphor. Other people maintain we should liberate computers from metaphors altogether. They argue that tomorrow's systems should be so natural and easy to use that they behave like the people, institutions, and objects we encounter every day. You just go out there and use them as naturally as you interact with people and things in the real world today. This is a seductive but unrealistic idea. Just to interact with other people you presume a level of intelligence on their part, which makes your interaction easy, but which, as we have seen, cannot be implemented by the machines.

Brain Chips

I cannot close this chapter on human interaction with machines without touching on a subject that is at the pinnacle of computer hype. Some time ago, when I had just given a talk on the Information Revolution, a young man approached me and said, "What I really want is a brain implant so that I can move massive amounts of information rapidly and painlessly into and out of my head."

"You mean so that you can download and upload information without going through the slow eyeball, mouth, and ear interfaces?" I asked.

"Yes. Isn't that a great idea?" he replied.

"No. It's a lousy idea," I said, "unless you are talking about sensor or effector chips," and went on to explain.

If you cannot hear or see, sensory implants are a godsend. Thousands of people who have inner-ear damage but a good auditory nerve have cochlear implants that restore hearing by converting sounds to electrical signals that excite the auditory nerve. Retinal implants, still in the early research stage, work in a similar fashion and may restore some sight to people who have a deficient retina but a healthy optical nerve. Experiments are also under way with people who have lost motor function; a chip, embedded in a muscle, detects the electrical signals from the brain telling it to flex. The chip transmits this information to a machine that will steer the person's wheelchair, giving her control where she had none before.

If sensor and effector implants are so great, why not place chips into the brain and perform even more spectacular feats? Every night, while asleep, you could download into your human memory entire sections of the *Encyclopaedia Britannica.* Or you could connect your brain chip to mine so we could intercommunicate our thoughts, directly and rapidly. Why even discuss spoken and visual communication with machines? Wow! Wow! Wow!

One basic objection is our ignorance. While we can channel simple electrical signals into and out of our body for sensor and effector functions, we don't know how to do so for more complex cognitive tasks. Where and how, within your brain, would a surgeon connect a chip's tentacles to communicate a simple command like "turn on the light," much less a concept like "freedom"? Scientists have been studying the brain for a long time, and they are still far from knowing how concepts are represented, let alone how to tap into them.

But let's be optimistic. Suppose after a while we crack the mystery of the mind and manage to connect chips to our brains so as to communicate our deepest thoughts. Wouldn't that make brain implants a great idea? Not quite.

Imagine that you and I and a couple of other people are successfully interconnected via brain chips. We might look cool with sockets in our heads. But we wouldn't be able to think; everybody's thoughts would be screaming for attention within our heads. We might then realize that some isolation among organisms is essential if they are to form a viable society. In humans, a balance between isolation and intercommunication is maintained by our seeing, hearing, speaking, and gesturing, whose slow speeds, compared to thinking, most likely represent the best that nature or God could do to preserve simultaneously the individual as well as society.

Not yet convinced? Then consider the threshold people must be willing to cross to violate the sanctity of their body. People with heart disease will consider a pacemaker implant or heart transplant only if there is overwhelming evidence that their life is in imminent danger. Few people would implant a chip into their brain for less than life-and-death reasons. We have wisely set a high threshold for tampering with the core of our being, not just because of fear, but because of natural, moral, and spiritual beliefs.

When I recently wrote about this in a magazine column I got mail from some people who were upset by what they perceived as my opposition to "technical progress." But as with "intelligent agents," this is an occasion where people are confusing a wish with reality. As director of one of the world's most forward-thinking research centers, I invite, even lobby for, research in areas we don't understand, including how the human brain works and how we may construct artificial brains. But I will not casually blur that which is imagined with that which is possible just to thrill the public with the shock of exhibitionist thinking or the promise of a utopia.

Let's get real. Let's find ways that help computers understand us through natural interaction, but also have a chance of working in the coming years, based on what we can see on the horizon. That's what human-centric computing is about.

Three
DO IT FOR ME
AUTOMATION

I use a great little electronic bulldozer in my office. When I arrived one morning, after my wife and I had decided to book an impromptu trip to Greece, I picked up the microphone and said, "Take us to Athens this weekend." It took me three seconds to issue this command.

The e-bulldozer, a computer program, knows that "us" is two people, that we like to travel business class, that we prefer aisle seats, and that the weekend comprises Friday, Saturday, and Sunday. It contacted the airline reservation system and negotiated with its computers by conveying the same query codes a human travel agent would type into forms on the reservation system's screens. The program started by asking if the kinds of seats we like would be available that coming Friday. If it didn't find them, it would have tried again for the two remaining days. But it succeeded. It then negotiated a few other things, back and forth, and found the lowest discount fares for the class of service we wanted. Five minutes later it had successfully booked the trip.

It usually takes the e-bulldozer between 3 and 13 minutes to complete this set of negotiations, because of delays in the airline computer and over the Internet, which the computers use to talk. Let's say that the 5 minutes it took to execute my 3-second instruction is typical. Five minutes is 300 seconds. These 300 seconds represent time saved from my life, or a travel agent's life, had either of us had to sit and execute all the operations my machine carried out on its own. Since 300 seconds is 100 times the 3 seconds I took to order this job, the improvement is 100-fold—a whopping 10,000 percent gain in human productivity! Not a bad electronic bulldozer. And not too far from a real earthmover, where a few ounces of pressure from human fingers on the right levers cause the steel bucket to lift thousands of pounds of earth.

Now you might say that my gain was only fleeting; I raised my productivity by 10,000 percent for only a few moments, and then was back to my old wasteful ways. True, if that's all I do with the computer. But if I were a travel agent for the U.S. State Department, I could sustain a tremendous, ongoing productivity gain compared to my old manual practices if I used this system throughout the day to book tickets for hundreds of travelers.

There was a time when people shoveled dirt with their muscles. When bulldozers came the workers threw away their shovels because they could move more earth, faster and more easily, with this automated tool. Today, even though we are seldom conscious of it, we are working harder than ever handling e-mail, browsing, and doing all sorts of "earthmoving" to keep our computers happy. It's high time we stop mentally shoveling and bring in the electronic bulldozers. That's what *automation* is all about—one of the most promising of the five key technologies that make up human-centric computing.

Automation moves beyond aiding our muscles, well above them, to replacing and reinforcing certain mechanistic actions of our brain. It does so with computer programs that control the appliances that serve us, but also with programs that manipulate information automatically and according to our wishes, where no physical entity budges, as in the case of my e-bulldozer.

We cannot achieve 10,000 percent productivity improvements across the board. But we should be able to raise human productivity by perhaps 300 percent during the 21st century. This gain will appear primarily in that broad category of human activity we call office work. Since it accounts for more than half of the world's industrial economy, this is a big deal, enough to fully justify the title "Information Revolution." A 300 percent productivity surge means we could do a year's worth of office tasks in just 4 months. We can take a nice long vacation and do everything else we've wanted to do but couldn't—for lack of time.

I have arrived at 300 percent by tracing the office work that takes place within the major sectors of the economy—finance, real estate, wholesale and retail trade, health care, education, the office side of manufacturing and transportation, government, and all kinds of services. I've sampled and estimated, though not in great detail, how much of what we do can be automated. This figure is roughly equivalent to the manufacturing productivity gain achieved in the last century of the Industrial Revolution.

The impressive gain will become a reality if we stick to the simple goal of human-centric automation: We want to prescribe to our machines how to handle automatically the information we care about, and how to control and coordinate our appliances—and then simply have them do these tasks for us, accurately, tirelessly, and repeatedly whenever the need arises.

The Ascent to Meaning: E-Forms

Once we tell our computer to do something for us, we turn our attention to other things, and it goes to work, interacting with other computers and appliances to get our job done. But for machines to work with one another, they must share certain common conventions—just like my e-bulldozer shares with the airline computer the conventions used to make flight reservations. This need for machines to understand what they are saying to one another marks a big change from the computing of the last 40 years, which was preoccupied with the structure, rather than the meaning, of information. The "meaning" that human-centered systems will understand is rudimentary compared with what humans share when communicating with other humans. Yet it can offer great utility to us.

The ascent to meaning is a pillar to the master plan for human-centric computing—and a challenge before systems designers and users to make it happen. It underlies not only automation but all the forces of human-centric computing. The ascent to meaning is a central piece of the business needed to finish the Unfinished Revolution.

A young inventory manager, working for a huge manufacturer of electric water heaters, is pretty sure that more replacement heating elements are sold when the weather gets very cold. She quietly tests her theory by collecting the prior year's daily sales information on these parts, from the computers at a few of the company's 600 outlets. She also tracks the average daily temperature in these outlet regions from a leading weather Web site. She looks at these sales-temperature pairs and her pulse quickens. There is a definite correlation. With care, she sets up a spreadsheet and carries out the calculation for all 600 outlets for the whole year. The result is stunning. Her model shows that by monitoring daily temperature, she can predict sales in each region to within 20 percent. This will let her safely reduce the heating-element inventory across all outlets by 60 percent, saving the company some $11 million a year. She takes the results to her boss, who is blown out of her mind. They quickly set up a new inventory control system for the replacement item, which is driven by automated daily temperature readings in the regions. After testing the scheme for two months, they cut it over, replacing the old fixed system they were using. Six months later at the company's annual holiday party, their CEO recognizes the inventive young woman's unique contribution, promotes her, gives her a 20 percent raise, and awards her a $30,000 bonus, following the company's reward rule of 3 percent of gross profit increase through innovation. The other employees applaud, while intensifying their dreams to parallel this feat.

Both this and the airline reservation example show the ascent to meaning. Automation is achieved when computers can "understand" what they are communicating to one another, so they can act on it. In the Athens example, my computer and the airline computer have a shared understanding of the date, number of seats, class of service, and origin and destination of a trip. They express these concepts with typed codes that were devised and agreed upon a long time ago by travel agents and airlines. The young inventory manager, for her part, revised the inventory control program so it would understand the codes and conventions established on the weather Web site for representing places and temperatures, as well as her own company's convention for representing daily sales at the computers of its 600 outlets.

We can think of these conventions used to communicate shared concepts as filling in mutually understood, prearranged forms on each participating computer. A form on the airline computer is filled in every time my computer sends it a query. And my computer gets its forms filled every time the airline computer responds. The "understanding" my computer and the airline computer show is in the actions they are each programmed to take as they process the information received in each form. Since the forms are communicated electronically rather than on paper, I call them electronic forms, or e-forms.

Depending on the adopted conventions, e-forms may require a special word before each entry, like "date," meaning that the entry will always refer to the travel date. Or the origin information might always be given first, followed by the destination, number of people, and so on. E-forms can be coded in many ways. The important notion is that there be an agreed-upon convention for conveying the requisite information from one machine to another, so that what is communicated is properly understood and acted upon by each participating machine.

Our familiarity with paper forms might lead us to believe that e-forms are filled exclusively by typing or by shipping text from one machine to another. Not so. In human-centric computing, e-forms can be, and often are, at the other end of a speech understanding system, and the entire purpose of the human-machine spoken dialogue is to fill them with the desired information.

Speech-driven e-forms coupled with automation can help us do much more by doing less. Consider how much time you would typically spend on a neighbor's phone when you call the local phone company to report a problem with your own home phone. First you are put on hold while you hear a recorded voice tell you how important your call really is. Then you get a human who asks all sorts of questions, types on his keyboard, checks repair schedules, asks for a supervisor to step in, and many minutes later tells you which half-day a technician might come to check your phone line. A speech-driven e-form would be far superior. Here's how such a no-wait exchange might sound between you and the phone company's service computer, after you have identified yourself.

> What's the problem?

I get no dial tone at my home phone.

> Okay, you get no dial tone. Please wait while we do a simple test on your line. (Pause.) Sorry, we can't fix the problem from here. Please give us a day and time our repair crew can visit you.

How about next Thursday at nine in the morning?

> Sorry, that slot is taken. Will an hour later be all right?

Yep, I can do that.

> Okay, confirming 10 A.M., plus or minus thirty minutes, Thursday, November 11, at 330 Cherry Hill Drive to repair a no-dial tone problem. Is that correct?

Yeah.

> Thanks for using Graham Bell Home Service. Have a nice day.

The entire conversation had but one objective—to fill in the phone company's e-form. At an even higher level, closer to what serves your interests best, you could say to your machine:

```
Please report a no-dial problem to the phone
company.
```

Your machine, preprogrammed to understand what you just said and to handle the phone company's e-form requests, would have carried out the above dialogue, pretending that it was you. Better still, the phone company's computer and your computer would have been already programmed, as part of your initial setup with the phone people, to report automatically any and all phone problems that the machines detected. All these are ways of addressing your service needs with e-forms through progressively greater automation.

E-forms work because the people and machines involved have established shared meanings. Control freaks immediately imagine the formation of worldwide dictionaries which would be used by everyone on the planet—gigantic taxonomies representing machine actions and transactions that help us automate everything in sight. This tendency to impose centralized procedures on inherently distributed systems, like the Web, exists a little in all of us; almost everyone who encounters the Web for the first time goes through a predictable stage of wishing someone would establish some organizing rules to make exploring and using the Web easier. Such centralized intervention is enticing, but it doesn't work, because it runs counter to human nature. People are as resistant to such universal shared conventions as they have been to the proposed use of Esperanto or Interlingua as a common, global language we would all speak and write. We simply do not want to bow to someone else's universal rules. If you haven't heard of these attempts, well, I've made my point.

It's not enough for only one pair of machines to share the same conventions. Broadly understood conventions about shared concepts are essential to automation. How can you and I and hundreds of millions of other people delegate to our machines a purchase from millions of different vendors, if all these machines don't have a shared understanding of "price," "product description," "credit card number," "delivery method," and so on? They all need it. Yet we just said the world will not, and practically speaking cannot, meet to agree on a universal dictionary of shared meanings. So how can this fundamental conflict be resolved?

Two ways look promising: gradual adaptation and the Semantic Web.

In gradual adaptation, the ascent toward meaning will begin with local agreements reached by people who belong to a common group or work in the same organization. Agreements will then arise across groups that share common interests and companies that belong to the same industry association. As buyers, sellers, and free exchangers automate transactions with one another over the Information Marketplace, they will establish shared terms that cut across the common interest groups and industry associations. The more routinely used terms will prevail and grow while the least used ones will dwindle and perish. And so the ascent to meaning will march on in a gradual, evolutionary way. This gradual spread of common terms will be occasionally punctuated by the "dictatorial" injection of "universal" terms from very large organizations whose wide reach will ensure widespread adoption. For example, the U.S. government could introduce a common vocabulary for carrying out the census, and for handling tax-related questions, while Interpol might introduce shared police terminology for thwarting cross-border crimes.

If you are reading between these lines, you'll have already deduced that achieving automation through shared conventions really means achieving human agreement among a group of participants. That is many times more difficult than the technology for handling these conventions. But it will happen, however gradually, because the payoff of greater productivity will spur people to do it.

Meaning on the Web: Metadata

Automation will involve computers and appliances communicating with one another on the Web. It therefore behooves the organizations concerned with the Web's evolution to introduce tools and procedures that make automation easy.

One such organization is the World Wide Web Consortium (W3C). Today this roundtable of some 450 companies, universities, and government research institutions establishes recommended technical guidelines for programmers to follow as they advance the state of Web software, so the Web can reach its full potential to serve people's needs. Such agreement is essential to prevent the Web from becoming fragmented into different dialects by different groups which would like to "own" it. Tim Berners-Lee, inventor of the World Wide Web, is director of W3C, which is headquartered at the MIT Lab for Computer Science (LCS), and hosted by LCS, France's INRIA, and Japan's Keio University.

One big W3C project concerns "metadata"—which means information about information. Its goal is to establish conventions and tools that help people from around the world with a common interest to represent agreed-upon meanings of information. Although you don't see them when you surf the Web, nearly every Web page carries "behind it" a few labels that describe the basic traits of that page—when it was created, who created it, the version of software used to create it, and the kind of information it contains (text, sound, photo). By Year 2000, most of this metadata was limited to these basic traits and was expressed in the familiar Web language HTML. The Web Consortium had already introduced additional languages with names like XML and RDF that can be used to represent e-forms and more complex descriptions of meaning, useful to automation. But this ascent to meaning had not yet taken off. Why?

One of the reasons is the difficulty people and organizations have in reaching agreements. Another less obvious but powerful reason is the economic makeup of the Web, which provides revenue to most Web sites through advertising. Automation is a major threat to advertising.

If your machine could go out on the Web and fetch what you need by examining a Web site's metadata, it would pass right by the ads on that site and return to your eyeballs only the information you want. You would never get to see the grand advertising that funds the site. With no funding, the site would go bye-bye.

So why would a Web site's owner tolerate the metadata that would make possible automation? Only if the revenue gained from automation were to offset the revenue lost from advertising. In time, the balance will shift toward automation, because there will be more Web businesses that generate revenue by delivering value to customers (rather than by selling ad space on their site), and because people want and will be willing to pay for the life-liberating benefits of computer automation, as they did with industrial automation. New business models may emerge that fix this problem in other ways, for example by splitting sites into pay and nonpay categories, like pay-TV and broadcast TV, or by splitting individual sites into these two components. Until we grow past the exhibitionist-voyeur stage of the Web, where advertising reigns supreme, automation will be fought. If we want to accelerate the onset of human-centric computing on the Web, we need to fight back, demanding automation and being willing to pay for it.

As automation comes to the Web, people will use the XML and RDF languages, which are already widely distributed and accepted, to create e-forms. Say, for example, that the various groups and organizations interested in rating videos agree on a set of three numbers, each on a scale of 1 to 5, that describe the violence, sex, and language of each movie. With a solid human agreement behind it, this metadata would ride on each video sent over the Web, on each tape rented from a store, on every review of the movie published online, and on every cable TV program guide. Every computer, television, and VCR would understand these ratings, and could be instructed, along with your Web browser, to ignore any video you or your children might come across with ratings, say, higher than 1 in violence, 2 in sexual content, and 3 in language.

People will also use XML and RDF to structure meanings in ways that go beyond simple text-oriented e-forms. Radiologists might agree to place a registration mark at the lower left-hand corner of every X ray. They could then type in or speak a comment about an anomaly on that X ray by clicking on it. The computer would dutifully register in the metadata the horizontal and vertical distance of this anomaly from the registration mark. The same convention might let them describe the anomaly through a shared vocabulary of standard medical terms, and add spoken or typed comments, at will. The metadata, expressed in XML, would consist of several items of text for patient identification, two numbers (horizontal and vertical distance), together with text or audio description and comment, for each anomaly. These bits of metadata would ride with each X ray. The U.S. Army, if it used this scheme for every person under its command, could rapidly scan the X rays of millions of current and past soldiers to locate the lucky people with a common anomaly for which a cure was just discovered. A doctor might then review the comments on the X ray of each individual, found by the machine, to ensure that each patient was indeed a viable candidate for the new treatment and would not be given false hopes. To understand the huge benefits of automation through metadata, imagine having to sieve through these millions of X rays manually and visually!

With hundreds of thousands of similar efforts under way by groups large and small in all walks of life and in all professions, human-centric automation will give us and our machines real power. Agreeing upon conventions for describing key bits of information about documents, images, audio recordings, videos, transactions, procedures, and processes can lead to massive productivity gains in every sector of the economy and in many personal activities. That's how the 300 percent productivity gain will be achieved. And because it involves so many different groups and organizations that must figure out how to do it well for their sectors, it will take the bulk of the new century to complete.

Keep in mind that no matter how useful such agreements within local groups might be, they will not be universal. Only a few will manage to escape the confines of the groups that created them. For example, a sliver of universally shared meanings might evolve for fundamental assertions and queries like "Yes," "No," "Price," "Address," and "Do you have X?" as people realize they can derive greater utility from their systems by adopting these basic meanings across groups. But this will take time. Ultimately, the sharing of meaning through gradual adaptation will serve primarily groups with shared interests.

For a more widespread sharing we'll need the Semantic Web—a new software capability that will extend today's World Wide Web and will surprisingly enable universally shared meanings without a universally agreed-upon dictionary of terms. I will describe it when I tackle individualized information access in the next chapter.

Bring Things under Control

Automation will help us do even more by doing less if it reaches beyond computer-to-computer conversations to the control and coordination of the physical devices around us.

I arrive exhausted at the hotel reception desk after a tough overnight flight. A long queue of equally tired people, framed by crimson-colored ropes, lies ahead, reminding me of the breadlines my mother and I endured in my native Greece during World War II. Fifteen minutes go by, before I am rewarded with an available clerk. She is comfortably seated, facing her computer. I am standing, facing her. She keys in my name, and after a pause tells me I have no reservation. I pull out my confirmation number and she types it in. "Ah," she explains, "that was for yesterday." Oops. I assume the mea culpa stance: I made a last-minute change and neither I nor my assistant notified the hotel. The nice lady tells me not to worry and asks for my credit card, which she enters into a different machine. She then begins to arduously scan with her eyeballs a half-dozen computer screens, looking for a room. We engage in the familiar ritual: She tries to palm off the second-floor room next to the noisy elevator, facing the street. I beg and argue for the exact opposite. She settles on a choice, and walks 15 lateral feet to enter the room num-

ber into another machine shared by all the clerks, which "prints" a mag-
netic key card. She returns, fills out by hand a little pocket card with my
room number and rate, and puts the key card in it. Weakened by the wait
I whisper: "At least it's a king-size bed, right?" "Yes sir," she says tri-
umphantly. I am done.

I rush to my room only to discover it reeks of smoke, even though my
reservation called for a nonsmoking room. The clerk apparently missed
the no-smoking request because of the reservation date mix-up. Hoping I
can change rooms before the anaphylactic shock sets in, I call the front
desk. After talking to an operator, I plead my case to another front-desk
clerk. He puts me on hold while he checks with housekeeping to see if a
recently released room is ready. Four minutes later he gets a "yes" and
assigns me to the room. Back down I go to get the new key card, then
back up to nirvana—28 minutes after I had stepped through the hotel
door. I undress for a shower and . . . in perfectly bad timing, the phone
rings. It's the operator who asks me if everything is all right. I suppress
the urge to tell her that her call made everything not all right, and ask
instead why she felt obliged to call. She tells me this is a routine check,
part of the new steps management is taking to offer quality service to the
hotel guests! I bid her gently good night, and try to suppress fantasies of
righteous punishment.

With or without my error, in a world of human-centric automation, my room registration could be done in under one minute. Upon
arrival, I swipe my credit card through a small device, which triggers
an automatic credit check and an immediate assignment of a room
matching my preferences, which had been stated when I made my
reservations. The computer would have caught my date error and
nonsmoking request, placing me immediately in the right room. A
key card would automatically be generated—inside an automatically
printed card envelope. Maids, upon cleaning a room, would have hit
a key code on the room's telephone that would tell the check-in computer which rooms were available, so it would already know where it
could assign me. And a single human clerk standing by would have
been available if I wanted to ask any questions.

With such a system in place, the waiting queue would have been 2 minutes, not 15. Avoiding the need for reassignment would have saved another 10 minutes. The hotel management would have saved labor by needing fewer check-in clerks. And if they got wiser, they could have saved even more labor by not calling all arriving guests . . . just as they put their foot in the shower. True service is offered, silently, at the moment it is needed, not at the server's convenience.

The speedup—and lessening of tension—that I would have experienced at the automated hotel would have been achieved through the interconnection of the check-in computer with physical devices: the credit card reader, credit checking system, key printer, envelope printer, and hotel phone system. The hotel's computer would have accelerated my check-in procedure by taking over, speeding up, and tying together through software several simpleminded interactions with physical devices that today are still done separately and manually.

The list of benefits from automating our physical environment is endless. Today you weigh yourself, then enter the weight in your computer diet program and basement treadmill. If these machines were interconnected, the computer and treadmill would be updated automatically when you stepped on the scale. Today the phone rings and you check the caller ID before answering it; if the phone were connected to your computer, it could filter out many unwanted calls automatically, without the phone ever ringing. The cameras and microphones feeding the computer in your aging mother's apartment would alert you if she fell down, or wasn't eating regular meals far better than any of today's alarm systems.

To derive these benefits, several prerequisites must be met. First, the devices must be able to communicate with computers. This is not as easy as it sounds. It presupposes that each appliance comes with a cable, or can broadcast a wireless signal, that can link it to a computer. This requires conventions that establish whether the appliance is a sensor, an actuator, or a combination of both; whether it speaks analog or digital; whether it sends data in a steady stream or in pulses;

and other electronic-level details. Even with hardware agreements in place, more must be done. The interconnected appliances must be told what to do by the computers that control them, through sequences of commands like "turn on," "louder," and "read room temperature." This requires software, and more standards, that let computers coordinate these back-and-forth exchanges with the appliances.

This nascent market was beginning to fill up in 2000 with languages and approaches like Sun's Jini; Sony's Havi; Microsoft and GE's SCP; IBM, Panasonic, and Honeywell's Home Plug'n'Play. Many more such approaches will appear, as users, together with computer system and appliance makers, smell the benefits of automation and go after it . . . with a vengeance.

A big technical challenge from a human-centric point of view is to establish an easy and natural way that lets people and computers "grab" the appliances they need with a plan that automates and integrates them. A plan can be expressed by an English-like recipe—a collection of commands, called a script. But locating the right appliances is a new and interesting problem. One approach already adopted by some manufacturers is to give every appliance an address on the Internet. As we have already remarked, that's not enough. It's like saying that everything in the world is at a designated physical location . . . and if you want it, go get it! A far better approach would involve some "ascent to meaning." Each appliance describes what it does, where it is located, and how it is controlled ("I am a camera in LCS Room 105, I speak protocol N21, and I am available for use by any LCS members"). A person or program might then ask, "Is there a camera near Room 106?" and on getting back the answer would proceed to use it.

That brings us right back to the familiar quest of securing broad agreements, this time for the "meanings" of exchanges between our computers and the appliances with which they communicate. To get there, we'll follow the same approach—gradual adaptation through human agreements and a broader, more universal approach through the Semantic Web.

Hundreds of Dumb Servants

Once computers can understand one another through e-forms, once they can ascribe meaning to information through metadata, and once they connect to and control the physical devices around us, we will start to feel the real power of human-centric automation. We will be able to stop shoveling with our eyes and brains, and create a whole bunch of procedures, each automating a specific task, to take over much of our tedious and repetitive work. Here is an example that assumes your home phone, cellular phone, and office phone have been interconnected to your computers. Seated with your feet up on your couch, you say aloud:

```
Shazam: From now on, if my daughter calls or
sends e-mail to my home or office, find me and
route the message to me.
```

First a word about "Shazam." Science fiction movies often use the attention-getting command "Computer!" to wake up a machine so it pays attention to what the human will say next. That's not very smart, because the word "computer" is part of everyday language. If I'm telling my wife at dinner about a thunderstorm that afternoon and say, "As soon as the lightning struck, the computer shut down," my home machine would take note and might faithfully shut down itself and all the other computers in the house! That's why I prefer the prompting word "Shazam," because it's unlikely to occur in normal conversation. Next time you have a spare minute, choose an uncommon verbal prompt you might like for yourself.

Awakened by "Shazam" and upon hearing you say "From now on," your computer would understand that it is being told to set up an automation script. It would construct the script, and check it by saying back to you:

```
Okay. From now on: (1) all phone calls from
caller ID 617 xxx xxxx or 781 yyy yyyy will be
```

routed to your active phone; (2) all e-mail
messages from sender alexandra@vvv will be forwarded to mld@zzz. Is this correct?

Upon my approval, the computer would convert this script to a simple little program that monitors the caller ID of all incoming phone calls to my various phones, and the sender's name of all incoming e-mail messages. It would then fire up this procedure, which would automatically test the script conditions I gave, for every call and e-mail I got. My computer can do this because it is connected to the right appliances (my phones) and to the right computers (the server that handles my e-mail), and can coordinate the information it receives from them based on my instructions (the procedure born from my verbal script).

Think of this little procedure as a dumb servant that doggedly performs the same narrow but useful task. Now imagine yourself after having spent quite a bit of time setting up a number of these scripts. You will be surrounded by procedures that automatically route your calls and e-mails to wherever you are, pay your bills, control the temperature of rooms in your home, alert you to news you care about, and much more. Collectively, all these "dumb" procedures serve you by being tuned individually to your various wishes and quirks.

Do many dumb servants an intelligent one make? Unfortunately, no. Up to a point, they're great. But too many procedures will sooner or later get into one another's way. For example, suppose I tell my system to never route calls to me for the first five hours after I arrive at an Asian hotel, following a transpacific flight, so I can sleep. What will the system do if my daughter calls at such a time?

It could tell me that two of my automation procedures are at conflict, and ask me to resolve it . . . but that would wake me up! Or it could have told me that a potential conflict was lurking, when I introduced the new script. Or I could periodically sieve through my procedures and add new ones to handle such conflicts—for example, "in a conflict, the daughter script should dominate all others"—thereby tuning my little coterie of helpers for consistency. Given these potential difficulties, I should always be able to turn an automation procedure off easily and rapidly. Human-centric automation should always be prepared to surrender control to the human, because a little bit of purported machine intelligence is often worse than abject machine stupidity.

Although not equivalent to an intelligent servant, the numerous automated procedures that will surround individuals at home, doctors, brokers, businesspeople, government employees, and many more participants in tomorrow's Information Marketplace will do much of their repetitive information work, and will go far beyond where we are today in helping these people do more by doing less.

Start the Ball Rolling

Since automation among computers begins with scripts developed by people and with agreements among people, we can prepare for human-centric automation without having to wait for new technology. We can then try the results with our current computer systems. If we want computers to "do it for us" we have to tell them what "it" is.

You can begin at your home or office by automating information exchanges or processes that you now carry out manually. Many commercial programs can help automate your electronic address and phone lists and calendars. But you can go further. I was able to cut by two-thirds the time I spend processing e-mail each day by combining with my Eudora e-mail handler a program called QuicKeys from CE Software that I have programmed to carry out sequences of actions, like sending a canned response to the recipient, a copy to my assistant, and trashing the original message—all with one click of my mouse on the right button.

You should also support the spread of metadata, and programs that can manipulate and translate it. Without this work you will not even be able to automatically add together the value of stocks you own in three separate portfolios on the Web, because the "total" from each of the three brokerages is on a different Web page, and is not identified as being a total by any metadata. Instead, it is simply painted on the screen. The total on the Web page for Portfolio A, for example, might appear as the fourth item from the left on Line 15. You could instruct a "screen scraper" program to automatically extract whatever it finds in that position. But if at some later time the brokerage house rearranges its Web page, as they often do, the total would change position and the scraper would pull the wrong information. Once brokerages start using XML and RDF so they can express the total as a piece of metadata and tag it with a descriptive name, like "total," then your computer could automatically calculate your overall portfolio value, daily, and have it ready for you at breakfast time.

Bigger automation gains will come once we stop being passive voyeurs on the Web and start forming those all-important agreements on meaning. If you are a manager, be a leader. Look at the exchanges of information that occur among the people in your department, or between the departments in your organization. Look at the information transactions between your customers and salespeople, between sales and manufacturing, between your subsidiaries and headquarters, and so on, lifting every stone you find. In each case ask: *"Could we gain time or quality or other benefits by automating this interaction?"* Chances are you can automate significantly, even using your existing computer systems. Once you know what you want to automate, the technical part is straightforward: You can introduce e-forms on widely used business programs, like Microsoft Office and Lotus Notes.

Since there are many information activities at most organizations, there is fertile ground. Going against you, however, will be people's resistance to reaching agreement across departments on what should be automated and on the conventions to be used. Overcoming that will require age-old management skills rather than new technology.

Even more rewarding are the payoffs that will come when your organization and others begin to automate your exchanges with one another. Here you must demonstrate the potential improvements, to build up interest among a few kindred souls within the other organizations who think as you do. They will comprise a most important core group that shares the same beliefs about what might be automated. Then convene a broader common interest group among the organizations that could benefit from such automation. Try to carry out a limited experiment among the organizations of your core group, to demonstrate the possibilities ahead. Avoid committees and standards groups at the beginning, because these bureaucracies invariably introduce long delays. Eventually, such standards coalitions will be required to establish due process in maintaining and upgrading agreed-upon conventions. But they will be more effective and move faster if they are preceded by a few specific, successful test cases informally agreed upon at the grassroots level.

A nice example of what can be achieved across organizations is bibliofind.com, a coalition of several hundred independent antique and rare-book sellers. They have created a shared search engine. When you log onto the Web site, you enter the title, category, price, publisher, or other information about a book you are interested in. You generally get the names of several independent bookstores in return; say, one each in Amsterdam, New York City, and Gilroy, California. The shared convention goes further; you can place orders from these bookstores in a single, online shopping cart. Bookfinder.com is another service, which searches several book-finder services like biblofind.com, abebooks.com, and usedbooks.com. Whether you visit bookfinder.com or any of the member services, you spend a couple of seconds to fill one or two lines of a standard e-form, which searches the inventory of all the participating bookstores. Imagine how long it would take you to locate, visit, and query each of them manually. You would never do it.

Bringing physical devices into the automation picture won't be so easy, at least until manufacturers make more appliances with special cables and plugs for intercommunication, and shared standards are agreed upon. This may happen quickly, though. Already, most automobiles have data ports that speed up the diagnosis of faults. Electronic appliances such as radios, music jukeboxes, alarm clocks, and washing machines are appearing with plugs and sockets for computer control. Market pressure may come from large organizations like airlines and hotel chains for machines with special sockets and exclusive standards.

Health care conglomerates that include hospitals and pharmacies could also drive demand. Doctors and druggists could automate prescriptions, saving precious dollars through greater efficiency and saving lives by avoiding errors and automatically locating suppliers of rare drugs. Hospitals, with proper privacy safeguards in place, could computerize patient records, and automate examination of those records to help researchers aggregate illnesses with common symptoms, in hopes of finding more effective therapies. The hundreds of companies making different items for the huge office supply chains could construct a marketplace on the Web where auctions for volume orders of each item are held automatically between buyers' and suppliers' computers. Such a system would lower costs through the supply pipeline, reducing prices to us.

There is no limit to the possibilities for automation. They are waiting to be discovered in every single business. Those who find them and act on them will be taking advantage of human-centric automation and will move ahead of their competitors. Hardly any of these activities carries the science fiction rush of an anthropomorphic robot that speaks with a tinny voice and cooks for you or sweeps your home. Never mind. The much greater excitement of human-centric automation lies in its off-loading human work from our brains and eyeballs, thereby helping us do a lot more by doing less. Go after it and get the automation ball rolling in your court!

Automation and Society

The question was sharp, the tone accusatory: "Won't the Internet and all these computer technologies eliminate our jobs?" The well-known politician furrowed his brow pensively and said smoothly: "No doubt, some old jobs will be eliminated, but new jobs will be created faster, ensuring increased employment. This is the way to the future."

I have heard exchanges like this in countless technical-political meetings I have attended in the United States, the European Union, and in every country I have visited that aspires to participate in the Information Revolution. The politicians differ but the message is from the same script. The response is fascinating, considering that Nobel laureate economists say we have no idea how advancing information technology may affect jobs.

If you think I am going to bash politicians for "lying," please guess again: Top politicians can't hedge, unless they want their flock to stand frozen before balanced but worthless assertions of the form "on one hand . . . while on the other hand. . . ." True leaders create a worthy vision, and move their constituencies in that direction regardless of initial consensus. And if they are extraordinary leaders, "when they are done, the people say 'Wonderful. We did it all by ourselves!' " (Lao-tzu, ca. 500 B.C.). The uniform attitude of the politicians I encounter stems from their intuitive belief that the new world of information is a worthy vision for their people. So if you hear these political statements, you know from where they spring. But what is really going to happen to jobs?

Economic principles state that increased employment will result only if the changes ahead cause demand to grow faster than productivity. We are quite sure that productivity will grow with the new technologies of information. But demand? No one knows. While we thus cannot peg what will happen to employment, there are a few things we can say about computer automation, drawing on our experience with industrial automation.

The Industrial Revolution's motors, electricity, and chemicals displaced laborers and craftspeople who worked with their muscles. In a generation or two they became bus drivers, jet pilots, managers of enterprises, and masters of new jobs. History will now be repeated. The automation of office work will displace certain kinds of office workers, and will, in time, create new jobs arising from the new technologies. As in the Industrial Revolution, the jobs lost to automation will be the ones that are repetitive and require little human common sense—tasks that a machine can be taught to do. The reporting of stock quotes, brokering of equities, and personal banking have already become largely automated. Look next for a similar transformation in computer system maintenance; preliminary screening of loan applications, insurance claims, and all kinds of office forms; hotel, travel, and car reservations; finding out about the weather and traffic; and exchanging basic information with government agencies.

How else might automation impact society? Some people maintain we will have to invent a form of mental jogging to parallel the physical jogging we now do to keep our bodies in shape. This may be necessary if automation is viewed as a human substitute. But if electronic bulldozers eliminate mental *shoveling,* they will leave us with more time for intelligent and creative thinking. Instead of wearing out our brains with mental drudgery, we will stimulate them with the kind of thinking we want to do . . . if we choose to do so.

A more sinister fear I encounter among my audiences concerns the possibility that if we delegate enough work to our automated servants, they may become sufficiently intelligent to match and control us. As I have noted, we have no basis for predicting that machine intelligence will increase to humanlike levels. We have not achieved any substantial gains in this direction, and don't seem to be on any promising track for doing so. That could change with a major discovery, at which point the prospects of the Information Revolution would change dramatically.

Ultimately, two human forces will determine what will happen with the automation of human functions: inquisitiveness and survival. Inquisitiveness will propel us toward greater automation, as we continue to invent new approaches that reduce our burden. But if we overdo this quest and surrender too much of our power to machines, our other age-old instinct—to prevail—will stand ready to obliterate anything, including automation, that threatens our survival. Out of these two opposing forces a new balance will emerge between the tasks we keep for ourselves and those we delegate to our machines. This new allocation of work among humans and machines will be defined by, and will define, the Information Age.

Four
GET ME WHAT I WANT
INDIVIDUALIZED INFORMATION ACCESS

Automation procedures would be even more impressive if we could lean back and simply tell them, "Get me the best information you can on X," and let them do the heavy lifting necessary to give us what we want. Unfortunately, today's information retrieval systems don't understand what we mean when we ask for something, and as they search, they don't understand what all the information they sieve through is about. They can only be scripted to look for matching words, which is what today's search engines do. To become human-centered, the systems that will find the information we are after must be able to discern something about the meaning of information. And since that is a difficult task, we must be prepared to augment their tireless mechanistic thrashing with a little help from our intelligent selves!

Finding the specific information that matters uniquely to us is vital to our personal and professional lives. It may be the changed arrival time of a loved one, the expected traffic delays on the commute to work, the results of a medical test, the price of a stock, a recently released report, the weather forecast, or the work we sell over the Net to a distant employer. Information is like money—rarely valuable in itself, deriving its value from the satisfaction of human wishes to which it leads. Properly informed, we are prepared to act upon our surrounding world. Having the information we need at our fingertips, when and where we need it, helps us do more by doing less.

We live in a world where the economic value of information is high and growing. However, finding the right information in this setting is formidable, because gauged by our individual needs and goals, most of the information out there is info-junk. (Consider how much money you would pay to avoid having unloaded on your front lawn the contents of 100 file cabinets, chosen at random from the United Nations' file banks.)

Today's much-heralded search engines that comb the Web can't help us much in sorting the jewels we want out of all this junk. They can only look at the structure of information. "Is this a text file or photo? If text, does it contain this magic 'keyword' my master has told me to look for? No—skip it. Yes—keep it." The better search engines have clever ways of narrowing the field while still bowing to structure. The Google search engine, for example, may find 2,000 Web pages containing a word you give it. It will then sort them according to the number of Web sites that point to each of the found pages, figuring this must be an indication of the pages' usefulness.

Another service, Blink.com, will store, free of charge, your Web bookmarks on its site, so you can use them from anywhere—your laptop, handheld PDA, or Web-savvy cell phone. While protecting the identities of its customers, the service compares the bookmarks you gave it with all the other bookmarks it has. It might find that 420 clients share the first bookmark on your list. The next time you open that bookmark, the service will recommend the most popular sites that the 419 other people share. The presumption is that these new sites may also be useful to you, since you already share similar interests through your first bookmark. Gleaning your personal interests is also what attracts Blink.com's advertisers; those whose goods you are most likely to buy will show up whenever you access the site.

All of these schemes, however, still rely on analyzing the structure of information. Has the file been accessed often? Does this search pattern match somebody else's? The results of these searches are still bulky and often do not contain what we are really after.

Very often, the information we need is somewhere on our own machines. And the content in which we need it may not match the context in which we wrote it or filed it away. We can't seem to remember how to describe it in ways the machine will understand. You may be looking for the letter you wrote to your landlord a year ago, and your system's "find" command yields no result to your search for "Jones," "landlord," or "apartment." You look in the folder marked "Correspondence." Still no luck. So you start opening every file and folder you have, until 15 minutes later you find it—it's entitled "Rent Letter" and it lives in your "Financial Info" folder. You are mad at yourself for not organizing your information better. Don't blame yourself too much. The semblance of order in computers, as in life, can be misleading. And forcing you to shove your letter in one bin of a hierarchic file structure isn't the best avenue for finding it later on.

What we really want our human-centric systems to do is to understand how we individually like to organize and describe information, and get us what we want, when we want it, whether it's on our machines or out on the Web. That is the goal of individualized information access—the third in our arsenal of human-centric technologies.

Organize or Search?

If we were omnipotent, starting from scratch, and interested in easy access, we would decree that all the information in the world should be organized according to one classification system—ours, of course, whatever it may be. Or, if we were public spirited, we might borrow from the famous Dewey decimal system used in libraries to organize information about books and magazines, crib from the Thomas Register that classifies industrial products, and repeat this process for every type of information we can imagine. We might also plagiarize from Aristotle, who worried about a taxonomy of all that surrounds us, or from hundreds of other ontologies that purport to organize the world. We'd then extend our system so it could represent emerging information, like Web pages. Finally, we would turn this scheme into Web metadata with its own universal vocabulary, which every individual and organization would be compelled to use.

If we could do this, finding what we want would become embarrassingly easy. But no such massively centralized categorization of information can succeed because of the age-old difficulty in reaching human agreements across a highly distributed world, where everyone has their own habits and ideas about how information should be organized . . . or left unorganized. Even if such a system were miraculously adopted, we'd continue to have trouble, because our notion of "proper" organization would change with time. And even if it didn't change, to make the scheme work, each of us would have to classify every new nugget of information we produce and store in our machines. Try impressing that discipline on people!

Still, the benefits of a front-end organization are so substantial that we should not reject it out of hand. Our human-centric technology for individualized information access will work better if we do a minimal amount of such work, which should involve no more effort than we expend today when we decide where to place a new file.

Since it's difficult for people to organize information when they first get it or create it, we may be tempted to devote all of our new technology to searching for the information we need, later, when we need it. But just because we can't expect much organizing from humans, it doesn't mean we can't get our machines to do some organizing on our behalf by having them inspect our information and the way we go about using it. The best strategy is to use technology for both purposes—to help organize *and* find information. Most important, we must ensure that our human centric approaches will be individualized, based on the premise that different people will make different decisions about how to organize or access the data that interests them.

We want the process of finding information to be as natural and familiar to people as possible. So we look for inspiration at the ways people typically get the information they need when computers are not available. First, we check our own desk drawers and bookshelves. Then we ask our friends, family members, and associates if they have anything useful, or if they know where else we might look. If these steps fail, we cast a bigger net covering as much of the world as we can reach, by looking in encyclopedias and reference books, going to libraries, consulting experts, and contacting institutions that might know what we are after.

This is the same approach our human-centered systems should take. Our machines will first check what they themselves might know, then go after the machines of our friends and associates, to the extent these people permit us to do so. If they still come up empty, they will roam the Information Marketplace, communicating with other machines as they try to discover information whose meaning is close to what we are after. And they may look back and forth among these three spheres for shared patterns and meanings.

Two current research projects illustrate what individualized information access could do for us in the near future. One involves looking for information on your own machines and the machines of your friends and associates. The other is for finding information on the Web. Both rely on the use of meaning. More approaches are under way, toward helping people find the unique information they need.

Discovering What Your Information Means

If machines are to organize your information, they will need to understand something about what your information means. Let's call the new breed of software that does this task "meaning processors." In a research project called Haystack, directed by David Karger and Lynn Andrea Stein at MIT, the meaning processors are free to roam over all your personal information. This includes every bit you touch, look, or enter on your computer—everything from draft documents and diagrams to e-mails you have sent and received, Web pages you have browsed, and chats in which you have participated. It also includes information about the many appliances you control with your systems, your spoken commands, and your automation routines.

There are two kinds of meaning processors in Haystack: extractors and observers. Extractors pull out key "header" information from a file. This could be the name of a database file or photo, the title and author of a Web page, or the sender's name, subject, and date of an e-mail. Extractors are programmed to recognize many different ways typically used by people to represent titles, authors, dates, and other such information. The extracted information is used to tag these documents so that they may be easily found later on, by title, author, date, and so on.

The observers track the frequency with which you use each piece of information you touch. If you access the same Web page of a brokerage service every day to check your stocks, that item will be tagged to signify that it is used frequently. Observers also note linkages among the pieces of information you use—for example, what you did when you dragged a file into a folder, and what you do immediately after browsing the brokerage Web page. If you go to another Web site that has a money market account of yours, the observer notes this activity as useful information. Most important, observers try to establish similarities among the information you look at. For example, if a Web site and a couple of your draft documents cite one another, or share a lot of words in common in their text or titles, chances are they deal with the same topic; the observers will create similarity links among them. Other observers watch your reactions to the results of queries, and give you the same results next time you make the same query, or the same kind of results if the new query is similar.

The extractors and observers work automatically, all the time. What they do to discover basic things about the meaning of your information is mechanistic, and does not require humanlike machine intelligence. But they stand ready to accept tips from you that will help them better organize the information according to its meaning. You may assign a descriptive tag to a document, or declare that two documents have similar meanings.

After the meaning processors have massaged your information for a while, files and linkages will end up being tagged with various meaning tags—similar to the Web's metadata. The first benefit to you will be an easier and more powerful ability to navigate manually through your own information, using these meaning tags as your guide. For example, you might tell your system to find all the files you've accumulated that relate to buying a car. Or you might browse one set of documents and follow the links to related documents, or clean house by browsing your least frequently accessed documents, deciding to relabel or archive some, and kill others.

In effect, Haystack generates a local Web that labels and connects your personal information, based on meaning tags. This becomes useful to you, because these connections are derived from your own actions and habits. You feel the power of all this preparatory organization when you want to find something. Rather than searching through file directories and e-mails yourself, you simply say or type "buy car." The system pulls up the information that has been tagged with "buy car." It also—and this is the important part—presents the links generated by the meaning processors and your tips, which thread together the "buy car" files with other documents and e-mail messages that are similar—something you simply don't have today in your computer. This makes it possible for you to pose questions to the system, such as "What information do I have on Toyota's passenger cars that relates to this e-mail my sister just sent me about a special Toyota offer?" Try asking this question in the "find file" program on your PC.

The payoff gets even more interesting. The Haystack system bundles all the links that interconnect similar information into a "bale." Each bale deals with a certain concept, though the system may not always know what that concept is. All the meaning tags, links, and the files they describe are grouped into a single bale already called "buy car," or soon to be called that, by you, if the system can't label it. Other bales might represent the concept "basement remodeling" or "family finances" or "music." Whenever you or your machine deal with a specific document, all the other files that relate to it will be easily reached.

Now suppose that your family members, friends, professional associates, and coworkers are all using a similar approach. Suppose also that after designating some of their information as strictly private, they have set up a standing permission in their machine to give you access to everything else, in exchange for the same privilege from you to them. Their observers and extractors have been sharing information with your observers and extractors. Your system's meaning tags and bales have been communicated to theirs and vice versa.

You and your system now have the added power to look for information in their data stores. You say or type "buy car" into your system and a link pops up to a file on your sister's machine, tagged "automobile purchase," which gives price information about the cars she investigated a year ago. Another link appears to a beautiful, short list that your neighbor, the car buff, has kept of the best car-broker Web sites. Its meaning tag reads "best car brokerages." Over time, your Haystack systems, working together, had established that "buy car," "automobile purchase," and "best car brokerages" share the same meaning.

The similarity link between your and your sister's bales was made explicitly by her a year ago, when she was buying her car. The other link between your and your neighbor's car-buying information was done automatically by the two systems two months ago, as they were searching for commonalities, and found the word "car" in the respective bale tags. To confirm their assumption of similarity, the two systems went further and compared the words in common among several files. Once established, the similarity among the three bales was made available to each system. The beauty of this approach is that while the systems established similarities among their files, the three users were able to preserve their own individualized way of describing the meaning of their information, and using it to find items of interest to them.

We have been talking so much about text, that we may forget there are other things we would like our computers to get for us. "I want an impressionist style painting of a woman sitting by the beach." Or, "Let me see titles of movies dealing with espionage in the Second World

War." These are legitimate human requests that we would like to address with the right human-centric technology. Unfortunately, today's meaning processors cannot pull out such a description by "looking" at a picture. And even though the topic is hot, and there is ongoing research for automatically classifying pictures by their visual content, we'll have to rely for a while on textual descriptions of this kind of sensory information. We can generate these text tags ourselves, or use automated options that label the image with the text of the Web page or e-mail that contains it.

With the meaning of information being so important to people, I expect the invention and development of many different techniques that will glean meaning from the information such computer procedures will examine. These new meaning processors should help individualized information access become better able to detect and link similar meanings, and hence become more useful to us. Let's understand, however, that these techniques will not free us entirely from doing some work ourselves, because of the limited intelligence that we can inject into software to extract meaning automatically.

The Semantic Web Conspiracy

Once your human-centric software has queried your machines and your friends' machines, the obvious next step is to query the rest of the planet. For the near term, this means the World Wide Web. The Web holds a vast store of potentially useful information, but getting at its meaning requires a different approach.

It was February 1, 1994, in Zurich, Switzerland, when I first met Tim Berners-Lee, the inventor of the then-young World Wide Web. He had kindly accepted my invitation to dinner and had endured the train ride from his residence in Geneva because he was looking for a good home for his "baby." He wanted an environment where the growing ranks of people writing software for the Web could meet and agree on technical matters that would help the Web grow, unfettered by special interests, so it could best serve all the people of our world. I wanted to meet Tim because I felt that the Web should be linked to LCS, and could provide valuable experience to our researchers who were designing information infrastructures. Things clicked between us, and after a few months Tim joined our lab, where we created the World Wide Web Consortium (W3C).

It was at that early dinner that I heard Tim's dream for the first time. His big hope was that as pieces of information became interconnected through the now-familiar blue Web links that we all click on with our mice, the growing web of interconnected information would gradually form a gigantic "brain." In Tim's dream, this new aggregate would become incredibly useful. It would start as a mammoth repository of human knowledge, but it would grow in usefulness as more people and machines threaded together the common meanings among that knowledge. It would help people find any and all information of interest. I reciprocated with my dream of the Information Marketplace, where millions of interconnected people and their computers would buy, sell, and freely exchange information and information services in a movement that would rival the Industrial Revolution in its societal impact. Over dinner we realized that our views were compatible and mutually friendly. We were eager to proceed. I remember thinking, at the time, that Tim's dream sounded far out. Almost a decade later, I still think a gigantic brain is out of reach, but we should be able to improve the usefulness of the Web to human purposes by injecting a healthy dose of meaning within it. LCS and W3C are working on this through a joint project we call the Semantic Web.

At this writing, the ideas on how to implement the Semantic Web are still being crystallized. It is not something separate from the World Wide Web. It is really adding a capability to the Web that can relate the meaning (the "semantics") of the information in its pages, pictures, and links. This new capability is central to our quest for human-centric computing and to the ascent from dealing solely with the structure of information to taking into account as much of its meaning as we can.

Let's say you've put off that decision to buy a new car until the year

2004. The doors are now rusting off your sedan and you can't delay any longer. Imagine that a hypothetical new piece of software, called Semantic Language, or SL (made up of XML, RDF, and the other leading-edge alphabet soup), has been added to the arsenal of Web languages and tools. SL was devised to make statements about the meaning of words, images, songs, and videos on a Web site (the familiar metadata), and about relationships among Web pages that have similar meanings.

By now, the Toyota folks use SL to describe their car specifications. The numbers they provide for each model's capabilities are tagged by SL descriptions like "model number," "horsepower," "weight," "price," and so forth. This is the familiar metadata tagging that has meaning to Toyota and a limited number of people who have taken the time to familiarize themselves with Toyota's Web site.

You are interested in cars that have high horsepower, a low price, and generous rear-seat headroom, because your kids are tall. You search the Toyota site to see if there is a model that matches your ideal car. But you also want to search the sites of the other automobile manufacturers, and you'd rather not do all these searches manually because it would take too long. Fortunately, most manufacturers have posted similar information on their sites, using SL. After all, it's been fairly easy to add the metadata tags of SL, since they didn't have to change the way their sites are organized. They also did so because they are proud of their products and want to be helpful to potential customers.

As you would expect, each manufacturer's site is organized differently, and sometimes the SL descriptions overlap or don't quite match. You wish the manufacturers had all gotten together and agreed on the same way to represent their cars' characteristics, but you know how unlikely such agreement would have been. What one maker calls "horsepower" in its SL assertions is called "power" by another, and "puissance" by the French automakers. "Rear-seat headroom" on one site appears as "backseat headroom" on other sites, and not at all on a third site. You wish you could find a list of synonyms that would match the like categories automatically. Fortunately, SL was created to cope with this need and provides the capability to establish these synonym maps, as the techies call them. The synonym links tell the computer, "Yes, the meaning of 'power' in this site is the same as the meaning of 'horsepower' in this other site, and 'rear-seat' has the same meaning as 'backseat.'" To your delight, and in response to your question, you discover a bunch of synonyms that Toyota has prepared for all the Japanese carmakers, and has posted them on the Toyota site, as well. With one click, you can compare the specifications for horsepower, price, and rear-seat headroom for all Japanese cars.

But that's no more than one sixth of the carmakers. What about the rest? Well, the U.S. National Automobile Dealers Association, in a fit of great service, has prepared an SL file that gives synonyms for the tags used by all U.S. manufacturers. (It wasn't necessary for the Automobile Dealers Association to do so. Anyone else could have produced this file, and as long as it was credible, it would still be useful.) A little note on the association's Web site informs you of where you can find other synonym maps. The General Motors site is listed as having the synonym maps between their models and those of their prime competitor, Toyota. Bingo! You found the missing link (so to speak). With it, and with the other two synonym maps you already have, you can widen your automatic search to include all U.S. and Japanese car manufacturers, bar none.

You are about to launch your query over this broader set of carmakers when you see the last note on the association's site: "AltaVista now offers comparison tables among all car manufacturers." You scold yourself for trying to do by yourself what was obviously one of the first things search services like AltaVista would go after. You quickly locate the table and you are impressed. Right there in front of you are all 65 of the world's car manufacturers with all their different car models, organized under one set of common tags. AltaVista picked the GM tags for that purpose. But it also makes the other tags visible, so if you prefer "puissance" you may search that way. Also, under the covers, AltaVista made sure that the synonyms it used are believable. It did so by using additional SL capabilities not discussed here. You begin to understand the gibberish you read a few months ago . . . something about the Semantic Web enabling the discovery of global meaning closures using pair-wise synonym links. In less than a second your system has scoured all the manufacturers and has presented you with three perfect models that meet your requirements.

Besides establishing synonym links between tags, SL also helps with conversions among related units of measurement. For headroom, it relates and automatically converts inches to centimeters. For power, it does the same thing between horsepower and kilowatts. And for price, it converts currencies to your currency, at the latest rate, which it gets from a site on currency conversions. The calculating ability of SL allows more complex relationships of meaning, too. The rear-seat headroom was not explicitly available in the Fiat specs. Never mind. The AltaVista engine found in its roaming an "approximate equivalence" prepared by an individual shopper, like yourself, which estimates the headroom by subtracting the ground clearance and an additional six inches from the car height. You can control whether these approximations are applied to your query by saying to AltaVista that you will only accept conversions made by people who were willing to digitally sign their contributions. SL goes further to establish and interrelate other kinds of relationships beyond synonyms, which give it even greater power.

SL can interrelate meanings using logic. Say an auto dealer 60 miles from your home in Atlanta happens to have one of the three models you are considering. You'd like to get an idea on a final price but would rather not make the hour-long drive just yet. The dealer has set up a private page on its Web site where price quotes and counteroffers can be negotiated online. The site's SL software decides who can join the negotiations, based on a set of conditions customers have to meet. Currently, it allows only visitors who have chatted online with a salesperson and have passed a credit check. SL makes this verification using rules expressed in SL logic. How do you know this is a legitimate auto dealer? The National Automobile Dealers Association provides a set of rules in SL logic for determining this. You can instruct your software to check the SL information provided by the dealer against these rules. When you finally buy that car, the dealer gives your system some SL code that goes into your automation subsystem. Later, your computer can check the manufacturer's Web site to see if any safety or recall notices have been issued for your model.

So in addition to providing synonyms, SL would also give people and machines the ability to relate meanings through arithmetic and logical calculations. Just to make things interesting, the different ways of calculating meanings are also right there on the Web. That way, if you have started an actual car dealership, you could access, using SL, the ways various dealerships like yours use to decide who can participate in online negotiations, and adopt or adapt one of them for your own use.

We must be careful in these musings here. A full logical "calculus" of meanings has been the dream of many scientists who perceived logic as a basis for emulating human intelligence, and who tried without success to convert this dream to practice. The idea is nevertheless fascinating, for even if it proved useful at a very modest level it would still extend the reach and utility of SL beyond synonyms to a wealth of derived relationships that interrelate meanings. Whether this becomes possible or not is not critical to developing individualized information access. Our human-centered systems can go far toward finding the information people need using only synonyms, much like people who speak different languages can go a long way by translating words from one language to the other.

This may sound like magic—establishing shared, universal meanings among a bunch of Web sites operated by organizations that do not share a common vocabulary. The Japanese, American, and other automakers, and the dealers association, for example, never met to agree on anything. How is it, then, that you ended up with a shared table comparing the specs of all the world's cars? Two things helped this happen: a shared human conceptual base, and a universally shared method.

All car manufacturers deal with more or less the same kinds of information, because they and their clients belong to the same *Homo sapiens* species and, hence, care about roughly the same things. There is an unwritten, shared human base of concepts like price, power, top speed, and so on, which is a natural consequence of people thinking alike about cars. It's hard to imagine that some manufacturer would report the number of different components in a car's transmission, instead of posting the price.

Enough small differences do exist among people and representation schemes, however, to make this shared conceptual base approximate rather than precise. That's where the universal method comes in. As long as car companies and third parties that compare car data use SL to provide tags and synonyms, comparisons among all data sharing the same meaning can be made. All it takes is a person or a machine like AltaVista's, in our example, to chase down all the synonym pairs and develop the famous "closure"—all the information tagged by a word and all its synonyms. The Semantic Language, SL, is the universally shared method. Please note that it is a method that lets people express in a uniform way their individual ways of organizing information; it's not a universal dictionary of meanings. People readily accept the former, especially if it is as useful as HTML has been, but have no use for the latter.

Who will create this ideal semantic language? The Web Consortium is halfway there, having already developed the XML and RDF languages. XML is a more powerful relative of HTML that was used to create most Web pages throughout the 1990s. RDF, which works with XML, describes metadata and is currently in use. It is being extended, as part of the Semantic Web project, to make possible synonyms. The other capabilities of SL that involve calculations of meaning are being researched and will likely make their debut as additions to these languages and as new languages. Other languages and conventions will surely come along toward the same goal.

Imagine the Semantic Web growing and growing, as more and more people and organizations thread SL links of meaning among almost every piece of information on the Web. Companies like Google, Yahoo!, and AltaVista that provide search services today will upgrade their searches so they are increasingly based on the meaning of information. Their huge computers will roam the Net, collecting all the synonym links they can find and organizing them into like-meaning clusters, so you can find the information you need far more easily than you can today.

As I close this section, I should reveal why I entitled it "The Semantic Web Conspiracy." Without realizing it, people, by creating local synonym links to serve their immediate purposes, will be building up a web of universally shared global meanings, which they would have never agreed to build in the first place. The conspiracy has a good chance of succeeding, because the forces that will propel people to establish local synonym links are the same ones that gave us the initial Web: Vendors will want their goods to be seen by the largest number of people, and will therefore establish every possible synonym link that will get customers to their site; and the millions of other people and organizations who create Web pages will do the same, since they will naturally want to be seen by even more people. As the links expand, individuals, hobby groups, professional and civic associations, companies, and government agencies will go after synonyms with a vengeance, so they can better share and find information, forever improving the clustering of information according to its meaning in the extended world of human-centric computing.

The Semantic Web will grow through the efforts of millions of people and organizations pursuing individual goals. And when it becomes big enough to encompass via its meaning a great deal of human activity, it will go a long way toward helping people do more by doing less.

A New Information Model

In automation procedures, my computer and the airline computer link certain pieces of information together, like "number of seats," so they can understand and act on shared concepts. Now we see that systems like Haystack link related concepts such as "buy car" on nearby machines. On the Web, SL will link metadata tags like "power," "horsepower," and "puissance" at different sites. And the best way to locate a physical device we care about will also involve some kind of meaning about what the device does; for example, "closest camera to Room 106."

In all these situations there is a single objective: to label information with its meaning and link together pieces of information that have the same meaning. Sometimes this is done with human agreements, like the airline conventions for reservations. Other links are created automatically by Haystack's meaning processors. Sometimes the linkage is introduced explicitly by you, when you tell your system that these two pieces of information refer to "basement remodeling," or that this device is a camera in Room 105. On the Web, the linkage will be done semiautomatically by organizations and search engines that group information by synonym.

When a single idea dominates so many different situations, it begs to become the model shared among them all. That's what will happen with the ascent to meaning in human-centered computers. We are laying the foundation for a new information model that will help us organize information on our computers, our physical devices, and the Web. This model is meaning oriented. Here's how I think about it.

The "meaning-oriented information model" is much like the familiar World Wide Web. Related text, images, sounds, videos, programs, software that controls devices, and other info nuggets are clumped together into a container called a hyperfile, which is like a Web site. In your personal system, one hyperfile might involve all the specs, photos, video clips, and e-mails you have acquired concerning a car model that interests you, suitably threaded to one another via the familiar blue links. You have three of these hyperfiles, one for each model you are considering. You thread these hyperfiles together with a new hyperlink you call "car models I like." This link is colored red, indicating it represents a shared meaning. The red links are threaded, just like the blue ones are. Click on the red link and you get the first car in your list, with all its information. Prominent on that new screen are the same words in red: "car models I like." Click on it again and you go to the second car. Keep doing this and you come back to where you started. Or jump back to "Home."

That's how your "file system" would look inside your new human-centered computer—a whole bunch of hyperfiles linked together with lots of red threads, entitled "basement remodeling," "budget," "personal e-mails," and so on. One important aspect of this information model is that a hyperfile can be threaded by different red links. For example, you may thread the "redo laundry room" hyperfile to the "basement remodeling" red link, and link it to the "basement remodeling cost" hyperfile through the "budget" red link. This sounds complex but is not, because you'll be able to get to any hyperfile you need from other hyperfiles related to it via many different meanings—something we seem to do with our brains as well. Perhaps you'll organize on your screen these red meaning links along a meaning-based, geographic metaphor of imaginary streets and towns, like "car city," "home repair building," "budget city," and "entertainment street."

Things get even more interesting, as meaning processors like Haystack's extractors and observers develop, by themselves, additional red links of shared meaning between some of your hyperfiles and those of your friends and associates. The results of Haystack queries, too, become hyperfiles and are linked to the rest of the information. Of course, at any time, you may create and name red links that make sense to you, if you are so moved. Meanwhile, the Semantic Web will be growing and all sorts of red links will appear on the Web, as synonym information gets clustered together by search engines and other interested parties. Many of the same old blue hyperlinks on which we click today will, in time, link similar meanings, and turn red, elevating the Web to a higher plateau of meaning orientation.

The new human-centered information model will truly become invaluable when you can take advantage of all the potential interrelationships among the red links within your system, between your system and those of your associates, and among these systems and the Web. That's when you'll gain a new benefit we'll call information triangulation. Suppose you want advice on repairing your car. Your system notifies you that the red link on your machine belonging to the concept "car repair" matches the red link called "car, fix" of your neighbor. Your system fetches this link, and your SL software discovers a red link to a public Web site that tells you all you need to know. You were able to use the triangle "you-him-Web" to rapidly and easily get the specific information you were after, from the big cruel world of infinite information.

Consider all the information people will be storing in the future, and all the red links created among people with common interests and within related groups. The opportunities for deriving help from this meaning-oriented information model are compelling.

For the last 30 years, our information model has been the familiar hierarchic system of files and folders. Meaning is not in this picture, except a tiny bit in the names you choose for your files. This system is called hierarchic because it requires each file and folder to reside inside exactly one parent folder. That restriction doesn't let you link information that is in different folders. Aliases and shortcuts were invented to help get around this problem, but they can only do so much as the patchwork they are. More recently, the Web brought us a different organization, defined by the blue hyperlinks. The hierarchic restriction is gone, since any piece of information can now point to any other piece. That heterarchic organization is closer to what human-centered computers need, but it's not quite there, because we need to point not just at anything that may be remotely related, but to information with a similar meaning. That's what the new red links do, exclusively.

After decades of bowing to the altar of structure, the time has come for a radical shift of our attention toward a meaning-oriented information model. It doesn't have to be exactly the way I described it here, as long as it becomes part of our human-centered systems and serves our need to organize and access information using its meaning. The meaning of information in the 21st century should become a central concern of builders and users alike, for it is the natural way people deal with information. Yet this most important aspect of human-centered systems—the ascent to meaning—will be gradual and imperfect, because our machines don't have enough intelligence to infer meanings the way we do, and because people are not prone to agree easily on shared conventions of meaning. But that shouldn't stop us. We have the technology on hand to begin this much needed improvement.

Call to Action

While creations like Haystack and the Semantic Web are being developed, we can go a long way toward improving the way we find information, using the equipment and software we already own. And in doing so, we will unwittingly be helping the transition to a broader, meaning-oriented information world.

Begin by asking yourself a key question: "What information out there is reliable, timely, and vital to me or my organization's purposes?" Perhaps you work at a small clinic and you have not yet made available to your doctors and nurses the many medical databases on illnesses, symptoms, and pharmacology. Or because of your clinic's location you get a lot of dermatology cases, but may not know that the University of Erlangen has DermIS, one of the world's most extensive dermatology databases, replete with pictures, symptoms, and case histories, all freely available on the Web. Go ahead and create links to these resources, if you think they will be useful to you and your associates. This may seem like a mundane activity, especially since all clinicians are bombarded daily with free offers to use various Web databases. But our objective with human-centered systems is not so much to excite as to pursue what is truly useful to people. It is precisely because the physicians in your organization are assaulted with all this information, that you or someone else must relentlessly sort through it to find sites that are accurate, timely, and applicable to your specific business.

It's to your advantage to find the best information out there. And that information is not standing still. New information is made available at an alarmingly fast rate. Explore this wild world and discover what is on the Web, on for-pay services, even on your own organization's databases that would help you achieve your goals. This search will take time and effort, but it is worth it. Asking peers and friendly souls in similar and related organizations for the information sources they use is a good way to begin. Look, too, at what competitors are doing. Also be sure to ask people within your company who may not be your first choice. You may be surprised at the distributed knowledge that exists near you about useful information. Finally, use your search engine of choice to find the sites and keywords that you think describe your key processes and goals. If the results are not exactly what you want, the programs will still suggest different keywords that will sharpen your inquiry. When you hone in on the best sources, create your own pointers to them. As you invest in all this hard work, remember that you are doing it to help yourself and your peers within your organization. The harder you work to find good information sources, the easier finding information will be for you and for them.

Once you have identified internal information and external Web sites that you think are truly useful to you and your associates, go ahead and turn these pointers into "red links." Since you don't yet have human-centered computers with their meaning-oriented models, this means that you will have to find alternative means to make these valuable pointers easily accessible to your people. At a minimum, this will call for posting Web addresses (URLs) and guidelines for getting at preferred sites. At a higher level of investment, you should build an internal Web site to introduce and maintain these precious links, for people who are not adept at hunting through the digital wilderness. Explore subscribing to useful services, such as Dow Jones Interactive, that not only offer newspaper and magazine articles of specific interest, but will also alert you whenever a key phrase you supply is mentioned in the media. One caveat: In individualizing information in this way, you may be tempted to offer your people too much. Don't. That's potentially as bad as offering too little, because people will get overloaded or frustrated and abandon your efforts.

Think of this task as managing pointers to useful information. Imagine that you are in the Library of Congress hallway, within a few hundred feet of most of the codified knowledge in the world. Does that help you find what you want? Of course not. What you need is someone to point at the information that might interest you. That is the role I am advocating you play within your organization, if you want to exploit the few, ever-changing gems of information that may truly help you and your company.

Another big question to ask yourself is: "What information within the organization is not currently on any machine but could be dramatically more helpful if it were computerized and made available to the other employees?" This is the very same question that information technology professionals have been asking for decades. But the world has changed since those early days of centralized data processing. Our systems and the information they store have become distributed. Now, instead of a few thousand people (the programmers of old) creating and organizing information, the task has become the province of a few hundred million people. Finding important new information to be computerized will help augment your organization's internal information for the day the hyperfiles and red links roll in. Meanwhile, you can get pretty close to that by ensuring that your new information systems are implemented in Web-like form on your organization's internal network.

The potential payoff is great. For example, hospitals and clinics are now studying the conversion of patient records from their voluminous paper files to machine form. The task is initially expensive and complex, because of conventions, privacy, and other issues. Yet it will revolutionize and dramatically improve health care by increasing accuracy, improving quality, and decreasing future costs. Interestingly, this task is considerably less expensive than it was when computer systems were centralized. At that time, the conversion to a new system at Children's Hospital in Boston was pegged at over $10 million, versus the $1.2 million it actually cost when it was carried out on a distributed Web-like basis. The writing and filling of medication orders has already been computerized in a few full-fledged health care delivery systems, like the BICS system at the Brigham and Women's hospital in Boston, with impressive benefits.

Has your vigilance about the information that could help you in this new regime risen correspondingly? From what I see in organizations I visit, it has to some extent, but nowhere near to the level it could. Chances are there is a lot you can do immediately, and on an ongoing basis, to bring information from inside and outside your organization to your associates. Encourage your associates to do the same, since the more widespread the effort, the greater the benefits. As you do so, your organizational information will become increasingly linked with other information out there, subject always to what you and others are willing to share. As millions of us do this, we will be enriching our information with meaning, and liberating the power of human-centered computers to get us the information we want.

Five
HELP US WORK TOGETHER
COLLABORATION

The European and American vehicle designers are meeting again, as they have done routinely for the last four months. They're part of a unique joint venture to create an exciting—some would say crazy—new product: a two-passenger, superlightweight car. Intended for cheap, local transportation, the car runs on a small 10 horsepower "lawn mower" engine and a couple of golf cart batteries. It also allows the driver and passenger to contribute to propulsion (and get some exercise) with a bicycle-like pedaling arrangement. The idea is to use whatever combination of power makes sense at any given time. The "car" is cute, shaped like an egg, barely 10 feet long, and weighs only 310 pounds. Driver and passenger sit in a reclined position.

The U.S. Motors engineers, in Detroit, are responsible for electrical and mechanical engineering, and putting the whole thing together. The Auto Italia people, in Milan, are responsible for the body design and interior. Also on the team is Delta, a Taiwanese firm that is handling the sophisticated electronics that control the vehicle.

Six Americans, early in their workday, and four Italians, at the end of theirs, have been meeting for an hour, each group in their company's "collaboration room." In each room, microphones hanging from the ceiling focus electronically, without moving, to pick up whoever is speaking, while suppressing the background noise. Video cameras high in the corners register the activity. Computers at each location manage this information, and communicate over the Net so the Detroit people see and hear their Italian counterparts on a 15-foot screen in the wall, and vice versa.

The meeting secretary is Max, a freelance engineer who is fluent in English, Italian, and Mandarin. The demand for his skills is high, so he can afford to work permanently from his lakefront house in New Hampshire. He sees and hears both groups via two 6-foot screens in his study. The computer in his basement is connected to the corporate computers at the other sites, through a high-speed Net service that Max pays for. The three machines are running a collab editor, a program designed for electronically coordinating such meetings. It records in audio and video selected fragments from what each person says and does, along with the reactions of the other members. It also keeps track of all the interactions the participants have with their machines as they call for slides, sketches, designs, and simulations. Max inserts spoken and typed keywords into the collab editor to mark major topics discussed and decisions taken. As the meeting begins, an Italian supervisor, driving back from Florence, calls in to join the meeting. The collab editor identifies his voice print and instantly admits him into the secure collab region—the shared electronic meeting space. His image is added to the video screens at each location.

A debate is heating up. Both camps are looking at the latest version of the three-dimensional design, e-mailed as an attachment from the central engineering team the previous evening. The Italian body designers want to change the shape of the plastic hood assembly to accommodate a three-frequency mobile antenna complex that handles two-way cell phone and high-speed network communications, but the American engineers are unhappy with the changes. The Yanks go on a quick, Semantic Web search for advanced antennas and find three others, but they are bigger and not as effective as the one they want to use. So they suggest a variation on the vehicle body by hand-sketching a different shape on their electronic whiteboard. The sketch appears on the Italians' whiteboard, and on their wall screen, as it is drawn, superimposed on the vehicle's blueprint design.

The Milanese ergonomics specialist is sure that the American proposal will reduce visibility through the front windshield. She picks up her aug-mented-reality wraparound eyeglasses and tells her American count-erpart to do the same, in slave mode. She then orders, in Italian, "Babushka! Adapt whiteboard sketch and give me driver's viewpoint." Meanwhile, the Italian supervisor tells his machine the message he is about to send should go only to his team. He dictates slowly as he drives: "Don't do it. Over budget." The message appears quietly as text on the Italians' confidential screen, which is shielded from the room's cameras. No one, including Max, is aware of this message—it's none of their business. The simulator, named for the Italian designer's Russian grand-mother, is crunching away, and as she moves her head she, in Milan, and her counterpart, in Detroit, see what the vehicle's driver would see out the windshield if the car were shaped in the suggested way. They easily agree that visibility would be reduced, unacceptably.

A young Italian, prompted by the budget message, gets the smart idea to make the change they all need, but only on the passenger side, leaving the driver with full visibility. The two teams congratulate the proud young man and agree this will solve their problem. They discuss a few more issues, then sign off. Max spends another 30 minutes summarizing the two-hour meeting in hyperfile form, using the collab editor. It enables him to index and summarize speakers, visuals, the whiteboard drawing, the blueprint, the 3-D simulation, and links to both company's design databases. He composes his text and spoken summaries in English and Mandarin, setting more keywords and cleaning up old ones to what was said and done. All three teams love Max's annotations, because they are direct and devoid of nonsense. He then joins his youngest son, just back from school, for a windsurfing escapade right off their dock.

Ten hours later, as they arrive for work at their office in the Taipei Technology Park, three Delta engineers assemble in their own collabora-tion room. They call up Max's summary in Mandarin, and look at the proposed design changes on their whiteboard. They also voice two issues they had raised over the phone to the Americans three days ago, and ask the collab editor to check if anything was said or shown that would resolve them. Thirty seconds later the Taiwanese computer displays on their wall screen the revised design that had been fully rendered in

Detroit after the earlier meeting. Two points spoken by the Italians, and keyworded by Max, are then spoken to the Taiwanese in Mandarin.

The three Taiwanese are delighted; in a couple of minutes they had found all they needed to know, without having participated in the two-hour meeting. Satisfied that their concerns have been answered, they are about to disband when the collab editor issues a constraint-violation alert. After the Detroit engineers had entered the design modification, a design checker program in Taiwan routinely ran the new specs against constraints the Taiwanese team had built into their own design database. It found a contradiction, which it reported to the collab editor: the new windshield curvature will be too sharp for the heads-up display—the ghostlike image of the vehicle's instruments projected on the windshield so a driver can see them without taking his eyes off the road. The engineers talk for a minute and agree they can make a simple adjustment to the projector that displays the image on the windshield. They voice a note, which the collab editor will append to its ongoing record of the design meetings.

One Taiwanese engineer is still uncomfortable about the shape change on the passenger side, but doesn't know why. He asks the collab editor to fetch the text, audio, and video that Max linked to the summary statement, "U.S. and Italian teams concluded that driver visibility would be okay if the shape changed just on the passenger side." The other two Taiwanese engineers, respectful of their colleague's intuition, which has been right all too often, join him in putting on their glasses to see exactly what the others had seen the previous day. "Aha!" the suspicious engineer shouts in his native tongue. He calls out with a wry smile: "And what are we going to do for Japan, England, and the other countries that drive on the left side of the road? Manufacture a second body shell with the molds reversed?" After a short, stunned silence the other two engineers voice enthusiastic sounds of approval—their friend had found an "obvious" point overlooked by the Westerners. They write an overly polite and formally registered technical memo about their insight. A few hours later, as the sun rises in New Hampshire, Max will translate it into English and Italian, phrasing it, as he always does, with the right cultural spin to minimize embarrassment.

The meeting secretary and the three engineering teams are all engaged in heavy information work. The gadgetry of microphone arrays, video cameras, augmented-reality glasses, monitors, simulators, and other programs merely support their intellectual efforts. The technological power comes from the collab editor, which lets Max apply his rare skills from home, and allows the teams to operate seamlessly across space and time. The collab editor weaves audio, video, design documents, summaries, and pending issues with the key threads of meaning that are important to the teams.

Without all this coordination, it would have taken days for the three teams to compare proposed design changes and resolve differences. It would have taken days more for the Taiwanese to sift through the melee and discover what mattered to them. The intuitions and insights that arose might not have risen at all if the participants weren't freed by the collaboration technology to concentrate on thinking and working together, rather than manipulating by hand the various messages, translations, drawings, and data. Without this technology, subsequent designs and prototypes would have been made, only to be torn apart and redone, in a mounting, costly fiasco, before somebody found the heads-up display contradiction and the "obvious" oversight about where drivers sit in different countries.

In time, this scenario will be repeated, with endless variations, in millions of different settings, including manufacturing, customer service, sales, medical care, publishing, finance, government, and a myriad of service industries. All of it is served by the same technology, the fourth pillar of human-centric computing, collaboration, that lets people work with each other across space and time.

I often hear that the Internet has matured because of the commercial transactions that it makes possible. Far from it! Human-centric collaboration will make possible activities that go much further. Information work will be the biggest beneficiary and will overshadow the purchase and sale of products, simply because it already overwhelms the latter in today's industrial economy.

The Challenge

The technologies of information have brought us a new possibility with gigantic repercussions: We can now electronically reach any one of a few hundred million people around the globe. That's a thousand times more people than we could reach when the automobile appeared, and a million times more people than we could reach on foot.

The challenge of human-centric collaboration is to convert this huge, machine-mediated human proximity into useful person-to-person and organization-to-organization collaborations. People should be able to work together synchronously—at the same time—or asynchronously—spanning different times—regardless of where they may be located. And they should be able to do so one-on-one, in small meetings or in large conference and theatrical settings.

For thousands of years, people have collaborated with other people synchronously and at the same location, in work teams, in class, at the market, in town meetings. Historically, people have also worked asynchronously, taking assignments home, or even across long distances, by conveying instructions to their collaborators via the mail. Imitating with human-centric collaboration the way we worked with each other before computers is important, because it preserves the naturalness of human actions and lets us carry forward with ease what we already know.

But duplicating the way we collaborated in the past is not enough. The new setting for collaborative information work involves people who must be able to communicate not only with each other, but also with their machines, across space and time, and in all possible combinations—people to people, people to machines, and machines to machines.

The forces of human-centric computing cover a lot of this new ground. Coworkers will be able to communicate with local and distant machines using natural human-machine interaction. They will order their machines to carry out work on their behalf by communicating with other computers, including devices and appliances. And they will be able to ask for and get the documents and other information they need. This leaves the coordination of human-to-human communication as the challenge that must be met by tomorrow's human-centric collaboration systems.

Synchronous human-to-human interactions are fairly straightforward. When people work together at the same time, regardless of where they may be located, they can easily make sense of what is being discussed, and with available technologies can hear and see each other and examine the same documents and artifacts. This kind of collaboration is smooth, because the participants are using their brains to understand what each other and their machines are "saying." But when collaborators are acting alone in their time slot, or are changing the composition of their teams over time, this is no longer possible. A way must be provided to carry the meaning of the various issues and tasks forward, along with the necessary documents and artifacts. Once again, the ascent to meaning becomes necessary, this time to bridge the delayed interactions among humans.

The coordination of synchronous and asynchronous collaboration will give rise to a new breed of collaboration systems. But even before these arrive, people can begin deriving the benefits of collaboration, using today's information technology. This is happening with e-mail, and in e-commerce, where changes have been swift. Bringing buyers and sellers close to each other, globally, has spawned a surge of novel business activities that have mesmerized the world with their potential. Hardly any of these transactions, however, involve direct negotiation between buyer and seller. The seller organizes descriptions of products or services, and the buyer, at some later time, chooses and buys some of these goods—hardly a collaboration, although strictly speaking, one that fits the definition of collaborative activity across space and time.

The vision of "frictionless capitalism," as Bill Gates has dubbed it, where buyers and sellers meet automatically without intermediaries, is rooted in this kind of minimal collaboration. It will happen for part of the purchase of standard goods over the Information Marketplace, like buying a book, or 1,000 shares of a stock. And it will stretch into other activities, like auctions. But the bulk of e-commerce, and most collaborative information work, will require intermediaries. If nothing else, we will need the go-betweens to intelligently sift the "diamonds" we care about from the ever-growing mountain of info-junk. People will also still want the substance, comfort, and trust of a human's involvement in transactions they care about. So even in e-commerce, which at first glance seems outside the realm of substantive human collaboration, we'll need new ways for people to interact directly across space and time.

Messages and Packages

At the turn of the century, the most widespread collaboration technology on the Internet was e-mail and its sidekick, the attached file, with an estimated flow of more than 1.5 billion messages per day. We can benefit greatly from this familiar technology, provided we make a serious midcourse correction.

E-mail messages include family exchanges, business letters, notifications, contracts, jointly authored documents, educational materials, software programs, government forms, medical forms, maintenance procedures, and much more. The immense popularity of this technology stems from its simplicity and its ability to bridge space asynchronously, while carrying with it "meaning" that is created and interpreted by human beings in their native languages.

Large companies like Dell Computer also use e-mail to provide service to their business customers. With the attached file, this natural process of written interaction becomes more powerful. The message "Here is my work," followed by an attached memo, song, picture, or a combination of these can go a long way toward handling human collaboration. Future e-mail will become more useful as it routinely, and incrementally, incorporates a growing mix of speech, diagrams, and video.

Besides its usefulness as an asynchronous medium of collaboration, e-mail is already growing to include synchronous exchanges for chatting and online collaboration. And who knows, in a few decades you may be able to don your bodysuit and skydive, then e-mail the experience to your friend, who will relive your descent through her own bodysuit and goggles. Perhaps she could even experience it with you in real time from the safety of her living room while you take the risk of jumping from a plane.

But such technological changes will hardly make e-mail better for people, because the medium has an inherent dark side: its propensity to sap our time. That's a problem that goes against the human-centric philosophy, and without correction it is guaranteed to get worse. We are headed for a 10-fold increase in received messages during this decade, as the number of interconnected people grows and as each person and organization increasingly uses e-mail. If you process incoming e-mails at 2 to 3 minutes per message, and you get around 20 messages per day, within 10 years your e-mail will require 8 hours of your daily attention, leaving no time for any other work. If, like most people, you treat e-mail as auxiliary to your main work, you can't let it exceed 10 to 20 percent of your time—an hour a day. Chances are you are headed for a serious overload.

What are you to do?

You can use a mixture of technology and human procedures to control it. Stripped of fancy descriptions, they amount to two options: birth control at the source, and euthanasia at the destination.

Human-centric e-mail behavior starts by avoiding the "look Ma" syndrome—sending messages and copies to show off, or to ensure that everyone remotely interested stays informed. Prolific e-mail authors should think of each message they send as an instrument that reduces the recipient's life by 2 to 3 minutes. They should send it only if they judge that the resultant effect justifies this cost. This may sound unreasonably harsh, especially since all human work involves invasions into other people's time that are generally accepted. But e-mail differs from face-to-face encounters, where everyone's time is equally taxed. If you take a mere 10 minutes to compose a message and send it to a list of 100 people, you will be consuming half a day of the recipients' collective lives.

E-mail birth control can also be achieved with office procedures, like an easy to use and socially accepted process for getting ones' name permanently removed from a mailing list, or a prize for the employee with the highest ratio of achievement to e-mail production.

Other steps can be taken at the recipient's end. Filters, built into mail-handling programs, can let people designate what messages to throw out or channel to other people automatically, based on the sender's name, topic, or other such information. You may place unsolicited e-mail in suspense mailboxes and have it reviewed by others, or by you, at a later time or not at all, or not until a second request from the same sender is received. With human-centric technology, Haystack extractors can obtain the links between messages—read and unread—so if you later find a message to be relevant, your Haystack will move the others closer in view.

Marketers should use the metadata capabilities of the Web to tag telemarketing information. This would help all of us control unsolicited messages, not only by getting rid of the ones we don't want, but also by letting through the ones we do want. It is unlikely that such an agreement would be reached spontaneously. This may be a situation where some appropriate governmental regulations could prove useful.

Even as it overloads us, e-mail is useful in helping work move forward, in assessing the pulse of an organization, and in receiving the opinions, suggestions, and ideas of fellow workers. To sustain these benefits, while increasing my own productivity, I have constructed an array of screen "push buttons," using the QuicKeys program mentioned earlier. When I click on a button, it inserts a preset note informing the recipient of my conclusion or question, forwards the annotated message to my assistant or the person who e-mailed me the message, and removes the mail from the incoming message list—all with one click. I have different buttons that say nicely, "Yes, I'll do it," "No," "You handle it," "Let's talk," and so on. I have been able to reply to enough messages via these push buttons to reduce my per-message average below one minute. No doubt, future e-mail packages will include such capabilities to help shrink our e-mail load. Of course, these techniques do not eliminate the fundamental problem. They just delay its onset.

The most crucial adjustment we can all make is to keep in mind and apply this simple yet crucial principle: Just because people have become interconnected, we have not acquired the automatic right to send a message to anyone we wish, nor the automatic obligation to respond to every message we receive. This shift in mind-set is the most important midcourse maneuver we can make to preserve the usefulness of future e-mail. Think of it as a human-centric procedure you, rather than your machines, need to adopt. If you like, you may freely repeat this aphorism at the foot of all your e-mail messages!

Ultimately, if e-mail overload becomes intolerable, the survival instinct will kick in and we'll trash everything in sight, as we should. After all, the principal role of information is not to be an end goal, but a means toward satisfying human needs. We should try to keep it that way.

As powerful as e-mail is, and even if we exercise it with the best human-centric attitude, it has serious limitations for substantive human collaboration. Consider all the natural collaboration activities you do in a day, like talking with another person, working with a colleague, meeting with half a dozen associates, or attending a talk. Imagine now that you carry out exactly the same activities, but with a twist—the only way you can interact with your coworkers is by passing notes to them and reviewing the notes they toss back to you. That would be much harder, slower, and less productive. This imaginary experiment makes vivid the gulf that lies between natural human collaboration and collaboration via e-mail. The same limitations apply to today's "groupware"—programs that help coordinate and track messages, documents, and meeting schedules among people. We need something better.

Collaboration Systems

To qualify as human-centric, tomorrow's collaboration systems will have to provide coworkers with the natural feel of ancient face-to-face encounters, while helping them draw on all the benefits of modern computers and communications. This translates into three essential system capabilities: making distant synchronous encounters as realistic as if they were held in the same location; carrying forward the "meanings" pivotal to continuity in asynchronous encounters; and coordinating these person-to-person encounters with the other human-centric technologies collaborators will use to speak to their machines, find information they need, and launch automated procedures.

Some of this design mentality has gone into the greatest collaboration system devised to date—the telephone. A synchronous phone conversation between two humans at a distance approximates the "naturalness" of a spoken conversation they would have face to face. That's why telephony has been so successful, even though the technology is old and "low tech" by today's standards. The fax machine gave people the ability to share images synchronously and asynchronously. Without such simple aids, office work would come to a halt. As we embark upon tomorrow's collaboration systems, we should remember that we don't have to bow before the latest whizbang technology to gain great human utility!

There's another lesson, too: The new human-centric collaboration technologies must reproduce faithfully the spoken word, and make possible the use of drawings, photos, and videos so that coworkers can show things to each other. Giving the participants the ability to see one another adds to the naturalness of the experience. Visual cues from the listener provide important feedback to the speaker without intruding on the spoken dialogue. But these visual accents are secondary, because mere vision cannot sustain a meeting at a distance, whereas the spoken word can.

In synchronous meetings that involve geographically scattered people, the technology should extend all the capabilities of a one-on-one collaboration to all the participants. It should also provide a means for each person to grab and relinquish control of the conversation, and shared documents, in an orderly way. In a synchronous theatrical or conference setting, the systems should enable participants to focus on the speaker and his or her slides, while letting them also participate with the audience in spontaneous comments, reactions, and questions. An important capability is to enable participants in other locales or at other times to "focus" visually and listen to any participant they wish, or pay attention to collaboration events taking place outside the mainstream presentation.

The technologies for achieving these capabilities are all within reach and are constantly improving. Spoken language and all sound can be transferred across distance with arbitrarily high fidelity. The one drawback is the awkward, fraction-of-a-second delay during conversations caused by the way the Internet ships its information over phone lines. So if you are a musician and your idea of synchronous collaboration is to play a piece with your fellow musicians across town, you will be disappointed, because the delays will keep you from playing with each other in time. These delays will be reduced as the Internet is gradually revised—a process that will be hastened by the need to use appliances and physical devices effectively, many of which depend critically on exact time coordination.

The electronic steering capability of gadgets like arrays of microphones that can focus electronically on speakers in distant rooms will improve, making it possible for several participants to use the same array to focus on different people and events. Voice quality will gradually get better, for speakers at stationary computers as well as people on the move using wireless portable devices. This means your voice will also be better understood by speech understanding systems at remote sites.

Showing collaborators documents and physical artifacts is also within reach. One problem is the cost of sufficiently fast communication "pipes" that can accommodate the high-speed flow of data needed to transmit video. Today's video transmissions over the Internet are small and grainy, but the deficiency will disappear in less than a decade, due to the hot competition between telephone, cable TV, wireless, and satellite companies vying to provide high-speed communication services, at low prices, to more and more people. As this sought-after "bandwidth" becomes more plentiful, off-the-shelf video cameras will be increasingly used to show participants to one another. Also within the next decade, a new breed of synthetic cameras will appear that can be steered electronically, like the microphone arrays, to focus on anyone and anything a participant wants to see.

Sharing documents like Web pages that reside inside a computer is fairly easy to achieve today. But working with your collaborators on a shared medium, like a map, blueprint, or hand-drawn diagram, is a bit more complex, though still attainable. One problem revolves around who can control and modify the shared medium and how such control is passed around, or grabbed, as different people make their contributions. A growing variety of whiteboards, camera-projector combinations, and other devices capable of detecting what you are writing have made their debut, but it is still technically difficult to register the various manual entries accurately against computer images, and project the results for all to see.

The synchronous collaboration technologies will also make possible the control and coordination of remote physical devices and appliances. An impressive early demonstration took place in April 2000 when a surgeon at Ohio State University Medical Center performed heart surgery on an ill man by manipulating tiny robotic cutting instruments from a computer console 20 feet from the operating table. Normally, for a surgeon to reach the heart, he must saw through the patient's sternum and pull back a sheath of muscle and ribs—just so that his own relatively large hands can get into the chest. But since the robotic fingers are much smaller than the human hand, they could be slipped right between the bones; far less cutting of the Ohio State patient was needed to perform the procedure, greatly reducing scarring and trauma, and shortening recovery time. In principle, the operation could be done from a continent away, and even by several specialists manipulating instruments from different locations. In practice, computers and networks will have to become much more reliable before they can be entrusted to support such life-and-death procedures. New activities, resulting from synchronous collaboration with the use of appliances and special devices, will no doubt emerge in many different forms and will surprise us with what they may contribute to our lives.

The timely presence of all these technologies will ensure that tomorrow's collaboration systems will handle fairly well synchronous work sessions. But what about asynchronous meetings? Ideally, asynchronous collaboration should mimic all the capabilities of synchronous collaboration, despite the delays and different players that may participate in successive meetings. That's a more difficult task, which will be tackled with the help of a new breed of collaborative software—a collab editor like Max used.

At minimum, collab editors will do for the collaboration process what text editors have done for document preparation. A collab editor would archive settled issues and track open ones. It would facilitate the recording and display of text, images, and video information, and record changes as the meetings takes place. It would admit or reject participants to the meeting. It would keep track of their interactions with their machines; for example, the running of simulations and the blending of sketches and design in the lightweight vehicle scenario. It would also capture, on cue or automatically, speech and video fragments from the discussion that are deemed important. The decisions on what is important enough to be recorded and linked would be made by each participant (by pushing a button), and sometimes by a meeting secretary like Max. The secretary would summarize in plain language the key issues discussed, and link summaries to the issues, the recorded audio-video fragments, and other information items, capturing the entire process.

All of this information would be recorded by the collab editor in the form of a hyperfile, where the text, sounds, images, slides, and video sequences, pointers to shared information, and shared computer data and programs used in a meeting would be linked to each other much like Web pages are linked, through a point-and-click of your mouse. The crucial elements in this hyperfile are the links that a participant would be able to follow later, to find out what was discussed concerning a particular issue, who spoke, what was seen, and so forth. The files would also give the participant in a subsequent meeting immediate access to everything that had gone on, which could be queried or updated. These links are the crucial threads that carry the essence of the meeting forward through time, to foster continued smooth collaboration. They connect all the items that have a shared meaning—through the familiar red links of our new information model.

Collab editors will also be useful in synchronous encounters because of the coordination they make possible among the participating coworkers and all their systems. Besides, today's synchronous meeting is likely to be part of a chain of asynchronous encounters, suggesting that consistent use of collab editors for all space-time collaborations would be good practice.

Collab editors together with their associated human procedures go beyond being technical support tools. They can have a significant socializing influence upon the people who use them. For example, who makes the decisions as to the rules of engagement for participants in a secure collab region? The boss, the participants themselves, or a computer program? Depending on how this is done, it could have a profound effect on how people feel about these get-togethers, which in turn could affect their motivation to meet, as well as their effectiveness as a group. At this stage of human-centric computing, we don't quite understand the significance and impact of this social dimension. I want to signal here that the human-centric tools we will use for collaboration have important additional dimensions that we should try to understand.

If you are not already doing so, you can begin exploiting the powerful force of working with other people through space and time, even before the collab editors, the collab hyperfiles, and the secure collab "spaces" make their appearances. E-mail is the easiest place to begin, and can take you far. Telephone discussions coupled with all parties looking at shared information like a document, a table, or a graph on their respective computers can be a simple next step. Such a text channel alongside the audio channel of a teleconference can also be used to communicate words, document names, and web addresses, which can be typed, rather than spoken or spelled out. And these text exchanges can be saved as part of the meeting record. Video teleconferencing can supply the added dimension of seeing and showing things. And notes from a human secretary can capture the vital meanings in asynchronous meetings. Today, these technologies are rudimentary, and they are neither integrated nor coordinated. And the video conferencing technologies are generally grainy and expensive. Nevertheless, these early tools can be helpful if you are willing to put up with some inconvenience and a bit of standing on your head to make them work.

Remember the utility of the phone and the fax machine. You don't have to wait until every home and office in the world is connected through glass fibers. Human-centric collaboration can begin today, with today's tools.

Information Work

Collaborations mediated by human-centric systems will not be effective in every instance where people need to work together. But they will be useful in many situations, especially compared with not collaborating at all, or wasting time waiting for an opportunity to meet physically. A huge amount of collaborative software will emerge in the near future, customized to professions and tasks. Medicine alone, with its various branches, will introduce numerous collab editors for doctors to confer around a hyperfile of your medical records, X rays, MRI scans, and laboratory results; for remotely monitoring and giving advice on surgical procedures; and for remote examination and diagnosis. Salespeople will devise their own breed of collaboration software, which will blend your bodily image with clothes you are thinking of buying, so you can see how you look in them, ask questions about them, and comment on desired changes.

In real estate, you will be able to see, from your kitchen table and with your broker's help, a bunch of different houses, viewing their exterior and interior under your control, and asking questions about price and the local schools. Certainly, thousands of realtors around the country are already putting homes on Web pages, but without the tours.

Lawyers will take depositions at a distance, and work with distributed clients and other lawyers to modify contracts using a collab hyperfile. In many businesses, especially the ones serving other businesses, the entire service they normally offer will be delivered as a distant, collaborative activity. An entire new industry will revolve around collaborative video games played by dispersed participants who can also see and hear each others' grimaces and screams. Since the suppliers of these games are accustomed to inexpensive solutions for mass markets, they may end up driving the evolution of inexpensive professional collaboration systems. The military and the intelligence services will collaborate across space and time for logistics, command and control, intelligence, and many other purposes, reducing expenses and personnel along the way. Dating services will take a new form, as couples explore their mutual interests through the appropriate collab software before meeting in person.

More advanced schemes will debut once distant collaboration becomes acceptable. Virtual- and augmented-reality displays will bring liveliness and the ability to immerse yourself in visual experiences—in medicine, real estate, training, machine maintenance, games, sex, and more. In time, ambitious software may also appear to guide, rather than simply record, these collaborative encounters. The opposite will also arise; in finance, government, and countless business and personal encounters, the collab hyperfile may be nothing more than a collection of plain, boring forms filled mostly with text and numbers. These tools, however exciting or lackluster they may seem, will contribute greatly toward finishing the Unfinished Revolution.

We will know that collaboration over the Information Marketplace has become useful when we suddenly realize we are using it routinely. Stop for a moment and think of the work you do, since chances are better than even that you are classified by economists as an information worker. How much of what you do when you work with others could be done with the approaches described above? And what about the rest of your organization?

Could your salespeople collaborate with customers, with each other, with the manufacturing folks? How about your vice presidents or subsidiary managers? Could some meetings that drag people in from around the world happen at a distance—or could you at least eliminate a few of them, by intermixing them with the new collaborations? Are there any services you now perform that you could buy at lower cost or higher quality from abroad? Could there be distant interactions with competitors—without breaking antitrust laws—for example, to set and maintain standards? Could collaboration technology allow you to relocate an office or plant, now situated on expensive urban property, to a more rural setting, which could be more affordable and serve as an added attraction for hard-to-find top executives? Could there even be some entirely new collaborations that improve your organizational performance?

As you consider these questions, formulate your answers against this test: If we pursue this approach, will we be able to do more by doing less? That's where your business savvy and technology will come together to determine how well your organization can leverage these human-centric techniques.

I also recommend that organizations form a small team of young-at-mind people who are Web savvy and have an intimate knowledge of the organization's inner workings. While committees can be deadly, killing a potential project even before it starts, this approach has worked wherever I have applied it, because the participants were chosen not as "representatives" of this or that department, but as passionate go-getters chomping at the bit to make a difference. Let these people "play" without any instruction other than to make suggestions on how the company might benefit from collaborations that span space and time. After the team has explored practical possibilities, it can bring ideas and prototypes forward. Decisions can then be made, based on common sense and good business practices, as to which should be pursued to the next stage—a limited scale experiment—and after that to full deployment.

You can also assess how collaboration technology can help you at a personal level. Participants might include family members, friends, faraway people you have lost touch with or always wanted to meet, and people with shared hobbies or interests. You might be able to interact with clinics, insurance companies, museums, your service station, educational institutions, or government agencies.

Ultimately, learning how to exploit the new collaboration forces on the Information Marketplace does not involve technological expertise. All that it requires is common sense, a knowledge of what is possible technically and what is desirable, and a willingness to be bold and creative while experimenting. Whether in offices or at home, a substantial part of the information work that people now do can be done remotely and asynchronously—and it will be done that way, whenever bridging distance and time to collaborate is more advantageous than working together in the same place and time.

Privacy

Despite its allure, collaboration raises important concerns about the privacy, security, and authentication of what is communicated over the world's networks.

The china at the electronic spy agency's dining room was exquisite, as was the meal. Ron Rivest, coinventor of the RSA approach to cryptography, and I were having lunch with the National Security Agency's director Bobby Ray Inman. We were trying to impress upon him that the forthcoming growth of the Information Marketplace would create severe privacy problems. We said the agency should extend the role of cryptography from ensuring secure communications for the U.S. government to protecting the privacy of U.S. citizens and organizations, using approaches like RSA. The admiral didn't agree; he thought our vision of a widely interconnected civilian world sounded like pie in the sky. Twenty-five years later, in April 1999, the Economist *proclaimed on its cover: "The End of Privacy."*

Underreaction then! Overreaction now!

No doubt, information technologies can be used to attack our privacy. But they can be used to protect it, too. For example, if everyone using the Internet did so under a scheme like RSA, creating and using their own pair of private and public cryptographic keys to encrypt their messages, we would end up with secure communications and files, not to mention the ability to digitally sign contracts and checks as effectively as we do now on paper. However, this high level of personal privacy technology would give criminals the ability to prevent the government from tapping a suspect's private data. It would also make anonymity more difficult, since the approach requires everyone to register their public key with an authority that can certify the person to whom a public key legally belongs. These issues can be resolved with existing technology and associated human procedures. We have technologies on hand to establish nearly any desired blend of personal privacy, anonymity, and governmental intervention.

Cryptographic techniques are integral to human-centric collaboration because they are needed to establish a secure collab "region" among several people who wish to work together over the "noisy" and less secure Internet. The objective is to protect their conversations and information flows, as well as their collab hyperfiles, from interception, corruption, malicious attacks, and plain old accidents.

To finish the Unfinished Revolution properly, our human-centered systems must be able to offer privacy when it is needed. We must be able to rapidly establish a secure collab region and then, just as rapidly, dismantle the region when we are finished. These collab regions must also be easy to join, by people who satisfy the rules of engagement for that particular session, while keeping out everyone else. The process is similar to that of a few workers in a huge, open floor full of noisy and occasionally snoopy people, who look for a quiet corner where they can carry out a critical discussion away from prying ears, and where they might easily be joined by a friend who should be included.

Establishing secure collab regions among members of a single organization is feasible because the participants are generally willing to use a single technological approach to that end. It is not as easy across organizations, because human agreement is far more difficult to reach.

Protecting collaboration is just one aspect of privacy in a world of high electronic proximity. Another aspect involves personal data about you and me. It is difficult to stop companies with which you do business from selling personal data you give them, or from corrupting it, or to stop them from gleaning your preferences and penchants by tracking Web sites you frequent. The problem is not a lack of technology. For example, a scheme called P3P, developed by the World Wide Web Consortium, places software in your browser and in the Web sites of vendors. You create a P3P personal profile on your machine, in which you specify the personal information you are willing to give away, along with what outsiders are allowed to do with it. Each vendor writes a similar script that identifies the personal information it requires and what it will or will not do with it. When your computer contacts a vendor's Web site, the two programs "handshake" prior to any transaction, and allow it to proceed only if both privacy declarations are satisfied. The same scheme can be used to establish absolute privacy policies—not just relative ones between consenting buyers and sellers. For example, governments could legislate privacy policies that would require all vendors to respect a certain minimum level of privacy in the P3P profile of every citizen.

These examples accurately suggest that we have enough technology around to provide nearly any level of privacy we want for collaboration and for the protection of personal information. But what do we want? In the United States, consumers treat privacy as a tradable commodity; we don't mind giving some of it away to get the goods and services we desire. Businesses that sell to consumers are enamored of this approach because they are moving away from mass marketing to one-on-one selling, and see their future hinging on their ability to build intimate knowledge about your, my, and everyone else's individual interests and habits.

To most non-Americans, however, privacy is an inalienable right, especially if it involves minors. The European Union, flexing its muscle, recently threatened to forbid its citizens from engaging in electronic commerce with organizations (read: the United States) that do not meet a minimum threshold of absolute privacy guarantees. The EU authorities then backed down and went to committee, as they and their American partners vowed to search for common ground. In February 1999, at the World Economic Forum in Davos, Switzerland, a few industrialists tried to establish a voluntary code, under which vendors would give you, on request, all personal information they had on you, explain what they planned to do with it, and correct it if asked. Adoption of this code seemed a small and achievable step, but it failed to pass. The American vendors saw it as an expensive proposition to implement, and a potential leak of their marketing secrets to competitors.

People disagree about the kind of privacy they want, and they don't seem serious enough, yet, about reaching agreements that could rectify the situation. At that same meeting in Davos, I almost fell out of my chair when several world leaders asked the technologists present to "go figure out a solution to the privacy problems you brought upon us!" This abrogation of what should be a central responsibility of legislators must stop, especially since the crucial missing dimension is human agreement. We must not surrender our privacy to the big lie of technological inevitability. In every part of the world, we must decide on what we consider more important—the government's desire to tap criminals' communications, and perhaps our own, or the citizens' right to privacy of their information. We must also decide what minimal privacy requirements we wish to legislate, especially concerning minors, and what we want to let float under free-market choices. And we must tackle anonymity and decide if and when we want to permit it. We must embark on these discussions in the world's national legislatures and within international organizations. Reaching agreement is a difficult but necessary and achievable goal. We have done it with passports, trade, airlines, and cross-border justice. Let's now do it with privacy.

Meanwhile, human-centric collaboration technology should be developed so it can provide an acceptable level of privacy to the people who use it. Since telephony has already established a level of privacy acceptable to most people, we can adopt it as a minimal, early goal of human-centric collaboration. We can then increase the stakes by requiring that this level be adjustable by the users, according to their needs.

More Social Consequences

Collaboration will affect the social fabric of our world more than any one of the new human-centric forces. This is because speech understanding, automation, personalized access to information, and customization affect primarily the individual, through human-to-machine or machine-to-machine interactions. Collaboration technology, on the other hand, changes human-to-human interactions, and therefore society. We must be sensitive to the social ramifications of distant collaboration. The issues are rich and complex. All I can reasonably do here is identify and describe some of them, in the briefest form.

On the economic front, the biggest social change resulting from collaboration will be the geographic redistribution of labor. Wealthy industrial nations will increasingly contract out information work to inexpensive information workers in developing nations. As happened in manufacturing, which was farmed out decades ago, information work will become a new facet of international trade, leading to tariffs and tolls, trade wars, and trade agreements. While today's information workers in industrially rich nations will view this migration with trepidation, the change will help raise the standard of living for information workers in the developing world. The programmers of Bangalore, India, who have been selling their software services to the industrial world, were making $10,000 a year at the turn of the century compared with $2,000 a year a decade earlier.

Taxation of collaborative and other commercial transactions on the Internet is a popular discussion topic. I often hear people say that these transactions shouldn't be taxed, because they are transnational, or because they will foster a new economy, or for any one of a dozen other reasons. But we have forecast that in a couple of decades, a quarter or more of the world's economic activity will take place over the Information Marketplace, and will include a large amount of information work. If these activities are not taxed, the collective tax revenues of the world will drop correspondingly. Nations will not tolerate such huge losses over the long haul. The question is not whether Internet business transactions will be taxed, but rather when and how.

Within a nation, the rise in information work at a distance will redistribute population away from urban centers toward less expensive, healthier family environments. A new class of people, the urban villager, will emerge. They will proffer their electronic services to the cosmopolitan centers of the world while physically living in rural towns, visiting the same shop owners and neighbors. Online buying and selling will help people acquire goods and services with greater convenience over the Net, and will hurt businesses and employees who now offer these products and services locally, at less competitive terms.

There may not be a net change in the number of jobs, however; as in the Industrial Revolution, some classes of jobs will wither, but new ones will arise. There will be a growing need for intermediaries for the purchase and sale of nonstandard goods—as mediators, information raters, and guides that will help all of us sort what we want from the mounting piles of global info-junk. All of this self-adjusting change will help some of the citizens of industrially wealthy nations and hurt others. History teaches that short-term remedies will be required for those affected, through social programs and human compassion. History also teaches that these forces will play their hand in the world's free economies and will move the equilibrium to a new distribution of labor, and a new set of skills, professions, and businesses.

The proliferation of information work through collaboration will raise questions about the ownership of information, an issue thrust into the spotlight in 2000 as several Web sites made possible the free sharing of music among millions of users. To avoid being dragged into the quagmire of copyright laws and other complications here, let's focus on the big picture. A carpenter uses his valuable human labor to build a chair. A doctor uses her valuable labor to build from scratch a database of symptoms, illnesses, and cures in her specialty. An artist who records a song also creates a potentially valuable result. If the chair maker is compensated for his skilled contribution by those who benefit from it, so should the doctor and the artist. Never mind that information can be easily copied. That doesn't change the value of the carpenter's, doctor's, or artist's work—just the price per copy. But when people feel free to access the information work of others without their consent, it's as if they are saying "I am now interconnected, so I have earned the right to steal a portion of your life for my own benefit." That's tantamount to conceding that human work is without value. We need not do so, nor do we have to reinvent our society. The problem can be resolved by continuing to let physical and information workers control whether they want to sell or give away their creations. All the technology we need is available in the form of cryptography, micropayments, and related approaches. We need only affirm with our will and with our laws that we continue to value human work above fads, technicalities, and subterfuges.

People also fear that the new, computer-mediated, human-to-human interaction will threaten human relations. It is true that one's list of local acquaintances could shorten as more distant acquaintances are added. But when it comes to deeper relationships like marriage and friendship, the primal forces that lie outside the Information Marketplace, such as trust and love, as well as hate, will remain dominant.

Human encounters across distance are unfortunately an ideal conduit for all sorts of criminal acts—predators going after your children; thieves trying to steal money electronically from your accounts; malicious offenders creating false information about you; spies prying into your affairs; terrorists attempting to hold executives, companies, or nations electronically hostage. The crimes these people commit are difficult to deal with because they involve cross-border violations that cannot be easily tracked, much less adjudicated. This is an area where government and law enforcement can and should intervene. As in the case of privacy, international agreements should be reached by the world's governments for handling information-related cross-border violations, in the same way that we now handle physical cross-border crimes and crimes in ambiguous jurisdictions like the high seas. We are already overdue in beginning these discussions.

And what of collaboration within the political arena? The hype suggests that large town-hall discussions could be held among thousands of people, and that there will be many more people-to-politician encounters in civil governance—even plebiscites to decide a majority of a nation's public issues. While enticing, these ideas won't work. Discussions among thousands of people are impossible because individuals can only cope with a small number of discussants and concepts. Also, too much citizen say in government could lead to chaotic rule by mass consensus, rather than by political leadership that sets a course. Representative government was invented precisely to handle these issues. It does not have to be reinvented, just because we have become electronically interconnected. The principal political effect of increased collaboration will be to further democracy, because collaboration will provide yet another channel for people to talk to one another, and because the major players of the new medium, who set the rules of engagement for everyone, are democratic nations. Any nation that wants to engage in information work or other transactions over the Information Marketplace, will be subjected to sizeable democratization pressures.

Distance Education

Let me end with a word about the world's most important collaborative human activity: education. Education is so vital because it defines future society. It's also the only force strong enough to close the expanding jaws of the rich-poor gap. It is natural for people to want to join new information technology with education. Unlike the Agrarian and Industrial Revolutions, which helped learners indirectly by feeding them, transporting them to school, and providing them with electricity, the Information Revolution helps directly, because it deals with the currency of knowledge: information.

Distance education has many faces. It can be used to teach literacy in Africa; provide industrial training and certification to health professionals, maintenance engineers, and other specialists; and offer courses to university and adult populations, as in the case of Britain's Open University. The biggest hope of all, however, is that it be used to interconnect young students with their teachers and peers, forming a new breed of educational communities that straddle space and time. Yet, despite the richness and promise of distance education, there is a dearth of responsible experimentation with its educational approaches. Partly, that's due to the difficulty of measuring objectively how effective a particular approach is. But there is also another reason—a wild-frontier mentality of hope and expectation that these new technologies are bound to help.

In the late 1990s, I attended a meeting where Benjamin Netanyahu, then prime minister of Israel, explained to a group of politicians and computer professionals how he wanted to provide a quarter million of his country's toddlers with interconnected computers. He said, however, he was having trouble funding the project. I turned the tables and asked him why he wanted to do this in the first place. He was stunned, since it should have been obvious—especially to an MIT technologist—that computers are good for learning.

Throughout the world, droves of politicians, led by those in the United States, are repeating the fashionable mantra that millions of children in thousands of schools must be interconnected. You can feel their rush: "Isn't it so responsible and modern to put an emerging technology to work toward the noblest of social goals: the education of our children?"

Not quite.

After 35 years of experimenting with computers in various aspects of learning, the jury is still out with respect to the central question, "Are computers truly effective in learning?" The evidence from numerous studies on whether computers improve the actual learning process is overwhelmingly . . . inconclusive.

Certainly, the promise is impressive. Simulators can help teach the kinetic and quantitative skills needed to drive, ski, swim, sail, even operate on humans. Computers can help learners write, compose music, generate designs, and create new objects. Speech understanding machines can be used as literacy tutors teaching adults who feel too embarrassed to fumble along in front of people, to read.

Collaboration at a distance can help teachers and students discuss homework on the Web, debate issues, examine problems, pursue joint projects, and get useful information from other people. At a more ambitious level, collaboration techniques can bridge schools that lack certain teaching specialties with schools that have the right people. Students in different countries can collect information on local customs and then assemble, share, and compare the results.

But *potential* does not equal results. Just to pick one statistic from a pile of evidence, U.S. high school students consistently rank from 12th to 18th, internationally, in physics and math abilities, whereas Asian students rank 1st. Yet U.S. students have far greater access to computers than their Asian counterparts. What are Asian educators doing, without technology, that American educators would do well to emulate? Another of the many reasons the jury is out is that learning depends critically on what human teachers do best—lighting a fire in a student's heart, nurturing a student, being a role model. None of these attributes are easily conveyed over the Information Marketplace.

So what should we do with this highest form of collaboration? I suggest the same answer I gave to Prime Minister Netanyahu, scaled up here to encompass the world: Let us interconnect students, and experiment with human-centric collaborative education, creatively and widely (in the hundreds of thousands to a few million students), but refrain from deploying it massively (in the hundreds of millions)—at least until the jury reaches some better conclusions. This won't make politicians shine as bright, but our children may shine brighter.

Six
ADAPT TO ME
CUSTOMIZATION

We have created a vision where human-centered computers converse with us, do our work, find the information we want, and help us work together. But how will your human-centric computer understand your spoken commands versus mine, automate your tasks rather than mine, find information on what you mean by "order supplies" and what I mean by "order supplies," and help you and me work together in a way that suits our personal and professional interests and is unique from the way other people may want to collaborate?

The answer is through customization. Customization is an essential part of our human-centric tool kit, because human beings and organizations vary widely in their interests, capabilities, styles, and goals. Computers must adapt to these differences if they are to help us finish the Information Revolution.

Customization on human-centered systems will be done the same way it is today—through applications programs. But there will be one big difference. What an application can and cannot do depends critically on the capabilities of a computer's underlying operating system. Human-centered operating systems will support speech and the human-centric technologies that bring computers close to the human level. This will make possible a radically new breed of applications that will be more capable of serving human needs, because they will be rooted in these new capabilities. This is why we have been so interested in the five human-centric forces. They are the new foundation that tomorrow's operating systems will expose, on which future applications will be built.

Customization will begin in earnest when you start using an application that comes loaded with specialized speech modules, automated procedures, individualized information access capabilities, specialized collaboration editors, and a great deal of customized software—all tailored to whatever specialty your application is offering. If you are a doctor, your medical application will come with speech modules that understand medical terms and medical talk, automation routines that operate medical devices and implement your clinic's procedures, information access tools that help you find medical data and papers, and collaboration tools for conferring with other physicians. If you are a banker, your application's speech modules will understand interest rates and financial talk, its automation routines will monitor financial indicators and alert you accordingly, its information access tools will help you find all kinds of banking data worldwide, and its collaboration tools will enable you to work with your clients.

That's only the beginning. The customization process will continue as you create your own added speech commands, automation procedures, and other wonderfully peculiar routines useful to you. This will make your system different than mine, even if we are both doctors, or bankers, and we use exactly the same medical or financial application.

Customization will extend to every corner of your daily life, too, as all this software becomes "nomadic." Nomadic software will flow wherever you are, whenever you need it, onto whatever hardware you are using—whether it's your laptop on the plane, your car computer, your handheld portable on the subway platform, or your wall computer in your office. Nomadic software will be particularly handy for software upgrades, and for use in wireless portables that can't possibly store in their miniature bodies all the system and applications programs you may instruct them to use.

A Growing Need

Carpenters, cobblers, upholsterers, and jewelers all use hammers. But a jeweler would crush a watch if he used the carpenter's hammer, and a carpenter would need a half hour to drive a single large nail with the jeweler's hammer. The customization of our tools has evolved over thousands of years, and it has served us admirably. Dentists, plumbers, and artists use a wide assortment of specialized physical tools that make them more productive and allow them to better service and delight people. In the bygone era of manual crafts, customized tools were so important that skilled workers made their own and signed them with pride.

How are we doing, by this age-old standard, when it comes to information tools? The knee-jerk answer is that today's application programs achieve the same thing, by adapting a general purpose computer to many specialties. That is only true to a degree; a word processing program allows a reporter to write a news story, and a drafting program allows an engineer to design a better paper clip, all on the same personal computer. But these application programs don't go far enough to adapt to individual needs. Novelists, poets, legal secretaries, doctors, insurance clerks, journalists, and elementary school students are all stuck with the same word processor.

I can hear my software developer friends' protest: "But we make the word processor powerful and versatile enough to fit a wide range of different needs." That's like saying, "We make one hammer versatile enough to fit the carpenter's and jeweler's needs." We passed this point a few hundred years ago. A single tool that tries to please everyone is like the classic Swiss army knife. Sure, the short knife can cut a twig, the tiny scissors can cut a string, and the stubby screwdriver can turn a half-inch screw in a piece of tin; but the knife can't cut a thick branch, the scissors can't cut canvas, and the screwdriver can't turn a four-inch bolt in a truck engine.

Software costs a lot to develop, and it's early in the Information Age, so we make do with generic applications designed like multipurpose shop tools for the homeowner. But anyone who has tried to use an all-in-one woodworking tool knows how difficult these machines are. You have to dismantle the old setting and set everything up for the new job. And then, as you try to do your current task, the settings and accessories reserved for other tasks inadvertently get in your way.

Word processors are a classic example. My wife was ready to kill the programmers of her word processor recently when, with no warning, each new paragraph she was creating in a letter was automatically numbered. That would have been great if she were making a list, but she wasn't. She tried everything she could think of to get rid of this "intelligent" (grrrrr) feature. But her 12 years of using computers were inadequate. She slavishly searched through all the menus for a clue as to how to override this annoyance—a procedure my colleagues and I also are forced to follow despite our 30-plus years of experience in designing and using computer systems. That confused her more by exposing her to a bewildering parade of features, with cute names that make sense to their designers but not to anyone else. She then resorted to the "intelligent helper" (grrrrr again) provided by the application. This software could not understand, in the terms she could express, what she wanted to do. The tremendous versatility built into the word processing program resulted in a system so complex that she could not use it.

This travesty is rampant. Big, clunky programs everywhere try to do a lot more than they should, in an effort to maximize their market. I'm sure you have a few you'd like to rage about. The result is confusion and very often the unjustified sense that you, the individual, are inadequate in your ability to use "modern" technology. We should all revolt and ask why people of our stature and ability should have so much trouble using a program that is touted by its maker as "user-friendly" (grrrrr, for the last time, bordering on violence).

I'll stop growling long enough to acknowledge with compassion and gratitude that much of what has been done with computers would not have been possible without these applications, however bloated and complex they may be. But technology keeps improving, and our human-centric objective calls for turning this gain into a new breed of customized applications that will go a lot further toward serving people's needs.

If you are a journalist in 2005, for example, you may be able to buy a new kind of word processor created with the advice of journalists who understand computers. This program would arrive electronically from a software service and would run on your machines. It would help you compose text with the right, yet minimal, editing features, while also letting you instantly and easily access the few newswire and video news services you care about. It would let you create a variety of automated procedures—for instance, one that calls your attention to fast-breaking news while you are editing, by monitoring the online news sources you gave it. The program would also let you access archived stories, photos, or videos from the data stores of your organization . . . and from a few other repositories you deem useful. It would do so with individualized information access routines that it had created by watching which sites you frequently visit. Because you treasure precision in your stories, you may not use the speech capabilities of your system to enter text. But you would frequently say things like "Go to Dow Jones now" or "What do the Brits have on this?" without taking your eyes off your newly crafted sentence, and the answer would appear in a box inset right in the paragraph you are working on. You might also say, "Find our nearest roving vehicle to this scene and route it there." Your word processor would also use the collaboration technology of your underlying operating system to bring up stories written by fellow journalists in your organization, and would let you hear or read the comments they had made "in the margin" that were never printed.

Imagining all these new capabilities, you might say that this is no longer just a word processor. You would be right by today's mindset. But you would be wrong by the human-centric yardstick of customization, because the new software would be a tool designed especially to help journalists put their words into stories—like the hammer that has been made especially for jewelers.

Customized hardware will take interesting forms, too. We see the trend emerging all around us. UPS and Hertz Rent-a-Car, for example, equip their roving employees with specialized electronic clipboards. Doctors and nurses use all sorts of customized gadgets. No doubt, tomorrow's novel devices will extend the capabilities of such tailored hardware. Hardware customization is done like software customization, but with an important difference: The capabilities of the tailored hardware should be widely used and error free, because after they have been "engraved" into silicon chips they cannot change. Arithmetic calculators, language translators, currency calculators, and label makers are typical examples today. For certain information tasks, these devices are preferable to their software alternatives because they are cheaper, faster, and more reliable. When was the last time your calculator crashed?

Another version of computer customization uses computer-controlled robots to tailor physical products described by software instructions, as in Levi Strauss's recent experiment to manufacture individualized jeans. In this scheme, customers from all over the world entered several body measurements into an e-form on the Web and a special pair of pants was cut by the robots and assembled by hand. After a customer had become satisfied with the fit, she could reorder additional pairs of pants. Some customers felt that the second and subsequent pairs should cost them less, but the pricing didn't allow for that. Levi's didn't launch this approach commercially because of high cost. Instead, a version of the service was made available through kiosks in Levi's stores. The Levi's folks believe that the principal asset they have is knowing the preferences of their individual customers.

Levi's discovered that customizing pants is still too expensive a process, compared with selecting from a range of bulk-manufactured pants the ones that should be pushed on a particular customer, given his or her preferences. In information tailoring, too, additional costs are incurred to collate news and other information items into a package that would appeal to individual interests. At first glance it seems cheaper to cut and paste news to fit a template of interests than to "cut and paste" cloth to make a pair of pants. But it's still too early to make such pronouncements. Many more experiments and business models will have to be tried before we get a clear picture of the future of customized physical products. I expect that this practice eventually will be widely adopted, as machines become able to crank out tailored products at mass-production costs, and as people come to prefer these products over the uniform goods brought to us by the Industrial Age.

Regardless of what happens to physical products, the customization of information will be big. Here is one way it might begin in the hot area of personal marketing: The preferences of individuals would be characterized by sets of numbers indicating each person's interest in certain products and product characteristics. For example, it may be that your attraction to classical music CDs is 3, while mine is 250—meaning that your frequency of purchase is only 3 percent of the average amount of classical CDs purchased annually by music-loving consumers, whereas I buy two and a half times that average. Let's also imagine that I have a propensity toward wood products that is 3 times greater than the average and 6 times greater than yours. If a company knows our preferences, and it is trying to sell a collection of classical CDs in a beautiful cherry wood case, its computers would quickly establish that it would be wise to pitch it to me and not to you. As a prospective customer, I offer a larger profit potential. And avoiding you allows them to further increase their profit by reducing wasteful expenses.

Imagine a company that accumulates a few hundred such preference "dimensions" for each of 500 million people around the world. This company would be an invaluable source of sales advice, which it could provide for a fee to other companies wishing to sell products and services, without ever disclosing how it arrives at its recommended lists of likely customers. A sophisticated marketing calculus will surely emerge on the Information Marketplace, one that will go well beyond such simple numeric schemes for combining individual preferences along several dimensions to determine how a particular new product would fare against potential buyers, and therefore in the marketplace. Insurance providers, finance houses, and basically all businesses that cater to individuals will go out of their way to customize their products and services, because that will differentiate them from the pack and increase their revenues. Look for a dramatic shift of business approaches toward customization, culminating in the ultimate marketing-oriented tailor-fitting—the management of sustained lifetime relationships with individual customers. The increase in market savvy through customization may well turn out to be as big a step as the introduction of demographics-based marketing in America in the early 1900s.

This kind of customization will affect people's privacy and will cause privacy policies to evolve. While the debates on privacy go on, more organizations are likely to discover and follow the practices of Amazon and Yahoo!, companies that pioneered the collection of personal information so they could learn as much as possible about their customers' preferences. These companies realized early on that to succeed they had to keep the information secret, and earn their customers' trust. I suspect that this kind of practice will prevail, worldwide, because it is responsive to people's privacy concerns and easy to implement.

Human-centric customization will also improve a consumer's ability to find what he or she really wants, by exploiting the cross-threading of products and their characteristics with semantic Web "red" links. Individualized information access and automation capabilities will be particularly useful in these quests. They will alert you when appropriate new offerings surface, based on a match of a product's features and your designated preferences.

Entertainment will be another big beneficiary of customization. Imagine being able to narrow down the existing stockpile of all 50,000 movies ever made, and place automatic alerts to "watch" new titles that are released, in order to arrive at the "perfect" choices that pique your fancy . . . which you can then rent electronically from your easy chair.

The terrain for customization will be even bigger in health care, finance, government, law, and a wealth of other services, simply because these activities overshadow in economic might, retail trade, and entertainment. Add to these the customization of business-to-business services and the prospects become huge.

Pushing the OS Upward

To take full advantage of this potential, we must equip tomorrow's computer systems with human-centric customization tools that people and applications can use easily and productively. The clear place to do so is with operating systems.

How does an operating system support and influence applications? The "color wheel" is a simple example. You may have noticed that your word processor, slide maker, spreadsheet, graphics program, and photo editor all have the capability to show a color wheel, from which you can pick one of many different colors for whatever you are doing. The basic ability to display the color wheel is built into the underlying operating system. All an application programmer has to do is literally "call" the color wheel, by inserting a tiny phrase in the application software that might read something like this: "color_choice = colorwheel." This "call statement" tells the machine to display the color wheel, wait for you to make a choice, and place the color you choose in a memory location called "color_choice," so the application may then use it for whatever purpose suits it.

Today's operating systems, such as Mac OS, Linux, and Windows, offer from a few hundred to a few thousand such calls to the application programs that run on them. These calls, taken together, form the operating system's applications interface, or API. The API doesn't stand still. New calls are introduced in new versions of an operating system to offer new capabilities. And because it is easy to make calls to these system routines, applications programmers are motivated to exploit the new capabilities of the API in new versions of their applications. To be sure, these new features may not always be useful, but they look good on the spec sheet and advertisements. Useful or not, the calls provided by an operating system penetrate all the applications that run on it, and give them a certain common character and feel, which makes us say, "This looks like a Mac application."

Unfortunately, in the four decades we've been using operating systems, their APIs have not risen much toward the human level. There is a myriad of low-level calls in today's operating systems—things like "close this window," "put this window in front of that window," "redraw this window's contents because the user moved the window hiding it" . . . just to pick on a handful of window management calls. Because applications reflect the underlying system capabilities, it is no wonder that when we are in the middle of some specialized activity, an application suddenly and stubbornly refuses to redraw or move a window.

I mention these issues not to put down the programming that has transpired so far, but to expose the intimate relationship between operating systems and applications, which we must exploit to make human-centric computing a reality. A great deal of credit is due to the people who have brought operating systems to where they are today. It is incredible how far application programmers have been able to go using the lowly machine-level capabilities provided by today's systems. But as heroic as these efforts have been, they have not been able to move applications significantly closer to people. The downward pull, exerted by operating systems toward what machines like, is just too powerful.

This pull has its roots in history. In the early days of computers, the limited technology available to software designers did not allow much of a reach into what users deemed natural and easy. New operating systems simply absorbed the differences in successive hardware models with calls that didn't change, so they could still support old applications. As time marched on, these habits set in and the level of the operating system became trapped close to the machine level. This left all user-related customization to the applications. To be fair, innovations in ease of use were made through the introduction of graphical user interfaces (GUIs) with their windows, icons, and menus. And even though these changes were modest, they evoked enthusiastic reactions from users, because they were so much easier than the old text-only approaches.

Such a revision must take place once again, but at a far more ambitious level, to bring applications closer to the level of what people want to do. The information technology terrain has changed sufficiently to warrant the design of such new operating systems. To succeed, these systems should be built from scratch, with a mind-set rooted in people's paramount need for greater ease of use and increased human productivity. In other words, they must include full support for the five basic human-centric forces, through a new and powerful set of calls to handle speech, automation, information access, collaboration, and customization. And they must support a new information model that is meaning oriented. The color wheel and many of the old calls will still be present in these systems, but will be hidden inside them, as subordinate internal commands that will be given to the lowest levels of a computer by the higher-level human-centric pieces.

The applications interface of computer operating systems must rise from its current machine orientation to a user orientation by exposing to users and applications alike the human-centric technologies. That is the most important foundation software makers can construct and application programmers can exploit to make human-centric computing a reality. Only then will application programs be freed from the low-level machine shackles of today's computers and soar to new plateaus of human utility.

Nomadic Software

When you walk into a colleague's office and ask to plug in your laptop, or walk into the local gym for an evening game of volleyball and ask to plug in your portable radio, the people present simply point to the wall outlet. Everyone, including you, accepts that this modest bit of electricity is essentially free. No one minds your stealing a few pennies worth of electrons. The wide availability and low cost of electric resources certainly lets everyone do more by doing less.

Imagine a day when computing resources can be treated with similar abandon. You walk into the conference room of your organization and approach the wall computer. The machine asks you to repeat a phrase it has randomly generated. You comply, and the conference room computer, recognizing you, adopts your info personality. This makes it possible for you to bring up information that you will use in a meeting you are convening in 10 minutes. Once the meeting is over, you clear the information in the wall unit, leaving it as empty as it was when you started.

This will come about, if people trust that the personalized information they bring to another person's computer will not be surreptitiously "lifted," and if computing resources become a nearly free good.

The first assumption has a good chance of coming about, since we possess the technology to offer essentially any degree of privacy and security we wish. The second assumption is questionable. In the last two decades of the 20th century, computer chips underwent a combined cost decrease and performance increase of 1 million percent! Yet, in the same period, the price of "personal" computers barely changed. The million percent gain went almost exclusively to greater performance and added bells and whistles. Still, within groups of associates and friends, within larger organizations, and within families, people use each others' machines without hesitation, making the notion of "free" hardware acceptable within their group, even at today's prices. With an ever-growing arsenal of portable wireless devices, whose price keeps dropping, this notion may spread further, outside and across tightly knit groups.

Customizing a hardware shell with your info personality so it can adapt to you is made possible by nomadic software. Your programs, hyperfiles of pictures, text, and video, automation scripts, speech modules, info access links, and collaboration practices and preferences go where they are needed when they are needed, taking over whatever empty hardware device is available. Nomadic software shifts our focus to what is important—ownership of information, rather than of devices. It couples individuals to the information they need, and uncouples that information from specific pieces of hardware.

You approach the empty skylit conference room of your company's branch office in Scottsdale, Arizona. The face recognizer and speech understanding software at the doorway identify you, as you tell it you will be leading the meeting that is about to start. In response, the wall computer system fetches from the company's servers a list of the automated procedures, info access bales, semantic links, and collaboration hyperfiles you may need during the ensuing meeting, as visiting colleagues from the Scottsdale, Kansas City, and Ottawa offices walk in. The system does not load all of the resources it has fetched into the wall machine, just the ones that past access by you suggests you might need. As you start the meeting and interact with the machine in front of you, it brings up the programs you need to do your work, as if you were in front of your own personal machine back in Kansas City.

The principal technical reason this rapid transfer of nomadic software will be possible is that the speed of organizational networks is expected to soar by 2010 to a hundred times or more what it was in 2000.

Other people in the room have the same needs as you. As they each speak in turn, the room's cameras and microphones determine who has the floor, and insert the new speaker's information personality into the wall computer, replacing the info personality of the previous speaker. At one point, while a Scottsdale manager is going on at length about a subject you know well, you become unhappy with this sequential arrangement. You need to check something on your personal system, without taking control of the shared machine. You reach into your pocket, pull out your little portable device, and pose your queries to it. This handy unit is not as fast or powerful as the big wall machine, but it can do a lot. It already knows your info personality and answers your queries well and fast. The Canadian associate next to you eyes what you are doing and gets jealous. She wants to check her own information stores. She winks at you and motions for the portable. You hand it over, and the little unit's inset camera and microphone quickly identify her. She then uses it, just like you did, except that now your piece of hardware is fully customized to her information world.

As your colleague beside you uses the handy portable, software flows back and forth among it, the wall machine, and other computers in Scottsdale, Kansas City, and Ottawa, using the company's high-speed network. If your seatmate had used a personal computer in her nearby temporary office the day before, her info personality might well come from that machine rather than from her office north of the border. The flow of information is governed by the system's desire to give her and you the best possible service by using plenty of wall machine power rather than your portable's power, and by getting nearby rather than distant information, if it is up to date . . . ensuring all along that your information is protected.

A similar situation holds when you use your portable machine on the road in Japan, where you went for some company business after the series of meetings in Scottsdale. Instead of other users' personalities coming into your machine, you now have many different pieces of your own information coming into and out of the portable as you ask it to do different tasks. Similar information swaps take place when new software or a software upgrade arrives from one of the nomadic software service companies you use, often without you being aware of the change.

All this software of yours will be distributed among your various machines and perhaps your organization's machines, depending on what you do. As you change your information, your human-centered systems, behind the scenes, will ensure that all the distributed versions follow suit. No doubt, it is technically possible for software to follow you around. The question from our human-centric perspective is whether this helps you, or your organization, do more by doing less.

Certainly, it is more convenient to use nomadic software than device-centered software; you don't have to lug around laptops or external disks to bring your information where you are. In a world of increasing mobility this is important. Another advantage is robustness; if your machine dies or malfunctions, or is replaced with a faster model, you don't have to spend time and effort reconstructing the information you had in the old machine. You simply get a different device and let the nomadic software fill it up with the right stuff. Yet another good reason to favor nomadic software is timeliness: You can always have the latest information up to the minute, and the latest upgrade of the software you are using. All of these attractions further the ability of tomorrow's systems to adapt to you, letting you conveniently have the information you want, when and where you need it. People will welcome the customizing ability of nomadic software.

Yet nomadic software is also controversial, because it raises the sensitive issue of who has "control" over the information you use.

To save money, your organization wants to buy all its hardware and software from a few vendors that offer favorable volume deals. Your company also wants to manage centrally certain information resources that are shared by all employees. These might include large-size printers, or "3-D printers" that build in 12 minutes plastic, three-dimensional architectural models, or central repositories of privileged data, like the company's patents. Your company also wants to upgrade en masse other shared, customized software that all employees use; for example, the special journalist editors that were developed by a large media company's programmers to give its 200 reporters an edge over the competition. Most large organizations also like centralized control because it improves the efficiency of software management, keeping costs down and ensuring a uniform degree of quality, reliability, and security for all employees. And even though they may not admit it, organizations also favor this approach because deep down they are still hierarchical animals, conditioned to the flow of power from a boss on top down an organizational chart to all the different levels of employees.

But newer management approaches are moving away from these centralized organizational habits. In the last 15 years of the 20th century, the most successful companies in the world discovered a great new truth that overturned Henry Ford's mass-production mind-set of telling employees exactly what to do, and treating them like a cost factor to be minimized. They concluded that their success was the result of giving employees the latitude to think for themselves and make their own best decisions in the interest of the organization's overall welfare. Belatedly, but admirably, they discovered that people count! Under the new mind-set, a shipping clerk would be encouraged to go out and buy custom hardware and software that need not be centrally approved, if he thinks it may save the company a bundle. That cannot happen easily in a company that regulates from a central information technology department all the software that people should use.

Encouraging grassroots, decentralized human power is what made possible the Web's growth to some 300 million participants in less than a decade. The Web made it easy for all these distributed "flowers" to bloom, each contributing and extracting what they want from the overall system. Anarchic as this approach may seem, it nevertheless has made big inroads in organizations, substantially redirecting their progress. This growing movement toward decentralization for buying, selling, and freely exchanging information has caught on enough that many organizations are building their own private, internal webs so that power can flow from the tentacles of an organization up, in full violation of the Church's innovation centuries ago—the hierarchy.

An even bigger force opposing the centralized distribution of software is the natural human desire to own rather than "rent" resources. Socialism has yielded to free-market capitalism. People don't like to use buses. And when they do, they dream of owning cars. Why should they use software that someone else chooses for them?

How might these opposing forces be resolved? On balance, the benefits of nomadic software customization, and of distributed human control, are so powerful that they have no alternative but to coexist. Here's how: The choice of the nomadic software that people use will become decentralized, while its distribution will continue to be centralized. This means that within your company you will pick the software you want, but you will also accept the distribution and customization scheme provided by your organization to get that software to your machines. In their personal lives, people will follow the same pattern, making their individual choices while accepting software distribution from tomorrow's software service organizations, software distribution services, and software clubs.

Customization rounds out the technologies of human-centric computing. We are now ready to apply them to serve our human needs.

Seven
APPLYING THE
NEW FORCES

Speech understanding helps you interact naturally with machines. Automation lets you control the physical devices you care about and create procedures that take over some of your information workload. Individualized information access gives you the power to locate and use, in your own way, the information you need, from your own stores and those of your associates and the wider world of the Web. Collaboration gives you the magical ability to work with other people across space and time. And customization helps you tailor your computer to your unique desires and specialties.

You could go a long way by using each of these new human-centric technologies individually. But the real power lies in combining them with each other, and ultimately in dovetailing them with the many human procedures we all engage in each day at home and at work. That's what pushes the capabilities of human-centric computing to a much higher plateau.

Here is an example of a short dialogue between you and your machine that combines two of the forces—speech and automation.

If Joe calls or e-mails, route his stuff to me, unless I'm on vacation.

```
From now on, all telephone calls and e-mail
messages from Joseph Bitdiddle will be routed
to you, unless you are on vacation. Is that
correct?
```

Yes.

```
Autoscript has been created. What do you want
to call it?
```

Joe's messages.

```
Okay. "Joe's messages" has now been fired up.
May I help you with something else?
```

Your machine uses its speech understanding capability to convert what you said to a scripting instruction aimed at the automation portion of your system. That part, in turn, creates an automated procedure that will monitor the headers of all incoming e-mails and the caller IDs of all incoming phone calls. If Joe e-mails or calls you, this procedure will reroute his message to your personal communication device, wherever you are.

Here are a few more examples of commands that exercise different combinations of the human-centric forces.

Get me last week's survey article on the flu. (speech, individual info access)

Show every participant the map Mary sent yesterday. (speech, individual info access, collaboration)

(Typed quietly while you are attending a meeting.) *Alert me if the building committee decides to budget more than $120,000 for the reading room renovation. (automation, collaboration)*

Please get Joe, Mary, John, and Ike now. It's urgent. (speech, collaboration)

If the nasty article by Jones appears anywhere, convene our watchdog group right away. (all)

Human-centered computer systems can do far more that just execute our commands, if they combine the five forces and bring them into our everyday routines. That's when modern technology, with all its rapid changes, will strive to match ancient humanity, which has not changed for thousands of years. And that's when bona fide ease of use and productivity will replace obfuscation and frustration. To fully appreciate the power human-centered computers will bring to each of us, let's consider a few detailed applications.

Health

Health care is one of the biggest potential beneficiaries of the Information Age. People are vitally interested in their own health, and want to have timely and easily understandable information. The medical community utilizes many physical devices that can be interconnected with computers to increase the accuracy and speed of innumerable procedures. And doctors are busy, mobile people who want to increase their own productivity in examining patients, diagnosing illnesses, carrying out medical procedures, and tending to meetings with other doctors and patients. Health care is thus ideally suited to three of the strong suits of human-centric computing: elevating technology closer to people, handling physical appliances well, and accommodating human mobility. Human-centric computing will create large, tangible benefits for doctors, patients, and health organizations—all the interested parties. As a result, its application in health care will rise dramatically during the next decade, provided patients can be liberated from the conservative inertia of healthcare institutions.

A growing computerization of medicine is scary to most people, because they fear that it will displace human care and human relationships and will threaten our privacy. Some such displacement has already taken place, and need not happen any more. Human-centric computing can bring big benefits to medicine without disturbing its human foundation. The changes will take place in the underbelly of medicine, where information is acquired, accessed, manipulated, and presented. In fact, by liberating medical people from routine and mundane tasks, the transformation may bring back some of the personalization that has been lost.

You go to your doctor for your annual exam. You also want to tell her about a pain you've been having in your lower back. Before meeting with her in the exam room, you enter the private premeasurement cubicle, where a nurse asks you to remove your shirt. A three-dimensional laser scans you and tells the nurse your chest size. She hands you a strange-looking but comfortable jacket. As you zip it up you feel the coldness of the electrodes that touch your chest at strategic places, and the pneumatic cuffs around your upper arm and chest that inflate automatically. You ask the nurse what happened to the wired harness that connected the jacket to the wall machine only six months ago. "It's all wireless now," she explains. "It's more comfortable." You joke that you feel human again—the old bundle of wires made you feel like your car, when the mechanics attach the diagnostic computer's test harness to it. The nurse explains that despite the changes, the jacket still takes your TPR (temperature, pulse rate, and respiration rate), performs a quick electrocardiogram, and measures your blood pressure. A hidden strain gauge built into the floor under your feet takes your weight, while an equally invisible laser scanner marks your height. The laser scanner that took your chest size also calculated the difference between your prior and current body measurements, and together with the electromyogram performed by the jacket to assess muscle tone, computed your overall fitness index.

The nurse reminds you that the next procedure will sting a bit. She produces a little, sterile thimble needle, pricks your index finger with it, and extracts a few droplets of blood. She then tells you to put your shirt back on, and asks you for a urine sample. You walk over to the bathroom antechamber and, as the door closes behind you, you say your name and repeat the sentence uttered by the machine: *"Mary had a little dinosaur."* You know the randomly selected sentence is there to block people who might want to fake their ID with someone else's prerecorded voice. Evidently your voice print matches your name and the voice print of the person wearing the jacket a minute ago. So the second door opens, letting you in. The toilet inside takes the sample without you doing anything special. Only the thrashing, scalding hot water jets that sterilize the toilet before and after you're done give a clue as to how this system works. This whole measurement session takes only eight minutes. You sit in the waiting room and relax.

Hardly a moment has passed and the nurse calls you to the doctor's office. You find your physician already examining the results of your preliminary tests. The results have also automatically been sent to your clinic's patient-record database, after the doctor's office computer filled in the right e-form for transmitting it and digitally signed and encrypted the electronic message, to ensure its authenticity and to prevent unauthorized people from seeing your private data.

You are impressed: On the large computer screen set in the wall behind your doctor's desk there's something that looks like a microscope slide of your blood, this just minutes after the sample was taken. You are also a bit anxious because there are all kinds of goodies swimming in that blood. Your doctor assures you that the *"swimmers"* are normal. But she is concerned about something you don't see—she says she sees a slight protein cast, which confirms what the urine sampler reported from its instant autoanalysis. She tells the computer, *"Get me the patient's Guardian Angel."* New information pops up on the screen. *"Aha!"* she exclaims. *"So you were sick three days ago. That explains it."*

Your Guardian Angel is your own personal medical monitoring program, which keeps an ongoing record of your medical history. Each person receives one at birth and keeps it until their death. Your Guardian Angel program resides in part in your personal information system, which you can access through your home computer or handheld device, and in part in your doctor's computer, which talks periodically to your personal system for, and in turn updates, your clinic's files.

You confirm that you did indeed have a bout of intestinal flu, which *you entered into your Guardian Angel. You ask your doctor if the protein cast is a worrisome development. She sees from your Guardian Angel screen that you had a low fever for three days and asks if you had nausea. You say no. After a few more questions she tells you that the protein cast and other symptoms suggest you might have had an incident of diverticulitis rather than the flu. Like your mother and father—and one in five people in your age group—you have a lot of little sacks in the walls of your intestines. Something probably got trapped in them, causing an infection. She tells her machine to perform a sedimentation-rate measurement on the blood sample you already gave, to check her hypothesis.*

All this discussion about blood samples reminds you that you have elected to decline analysis by the DNA microarray that measures the expression of some 30,000 genes in your white blood cells and can warn you about your risk factors for heart disease, cancer, and other major illnesses. You'd rather not know. But you are well aware that your brother Michael was greatly helped by this new capability. During a routine physical exam, his Guardian Angel noticed that his systolic blood pressure had been climbing by 2–3 mm for the last five years. With his consent for access and analysis, a drop of his blood was taken and in 5 minutes, the 200 most relevant genes for his increased blood pressure were sequenced. Then a Guardian Angel computation was performed using these sequences, Michael's medical history, and the history of his immediate family members. On the basis of the resultant "profile," the physician selected an antihypertensive medication that had been demonstrated to work on people matching Michael's profile. This profile information was uploaded on Michael's Guardian Angel and immediately expunged from the physician's and clinic's records, based on a standard agreement negotiated between the Guardian Angel and the genotype equipment.

Suddenly your doctor's system issues a low but distinctive shrill tone. It's the signal that she is wanted for an emergency consultation. She asks you to wait and puts on her earphones to ensure privacy. She then engages in a discussion. From what you can hear, and see by straining to look sideways at her screen, a surgeon in an operating room at some hospital is asking her questions about a patient of hers lying on the table.

An MRI scan superimposed on what appears to be a live view of the patient's abdomen fills the screen. She responds with a long and incomprehensible string of medical jargon and the discussion is over. You are a bit annoyed at the interruption, but you know it may be necessary on your behalf some day. So you applaud the process that enables physicians to hold a quick, emergency collaboration during an otherwise routine examination session.

During the brief interlude you remembered to tell your doctor about your back pain. You tell her you are worried about possibly having a herniated disk, and how you felt it most when you were doing some landscaping the prior weekend. You proudly tell her how you had done an automated search for clues on a medical program you have at home and some medical Web sites you reached, on which you entered your symptoms. She hits a key and a voice fragment from your Guardian Angel reports exactly how and where you felt the pain when it was at its worst. She smiles and asks you to lie on your back on the exam table, and pull your knees up to your chest. She then tells you to push outward with your feet while she resists. She probes your back a bit as you do so. She then tells you your pain can't possibly be from a herniated disk. She also gripes politely about how so much computer acculturation has made it easy for too many people to misdiagnose themselves. She says you probably strained a muscle in your back a week ago. It rings a bell: You were indeed building a stone wall in your yard. She goes on to voice a prescription for a painkiller you should take for the next three days. Her computer demands the doctor's electronic confirmation that what it understood is what she intended; it then sends the order to the local pharmacy, and to your Guardian Angel, which will vibrate in your pocket when the prescription is ready—and three times a day to remind you to take the pills.

Your doctor briefly examines you with the old-fashioned stethoscope while squeezing and pressing your abdomen in familiar ways that have not been automated. She then leans back and asks you how you are really doing. The leisurely discussion that ensues is the primary benefit of all this computerization. It enables ample eyeball-to-eyeball interaction between you and your doctor—and invaluable diagnostic asset at the heart of medicine that almost got wiped out in the accountant-dominated medical environment of the late 1990s. Soon your exam is over. You marvel that while it has been more thorough and satisfying than the old physical exams, it has taken just about the same time. As you leave you pass the reception desk. You place your little handheld unit on the counter. It instantly downloads the test results using wireless technology, updating itself (and, later, your personal data at your home system) with the whole visit's data and directives.

As you drive from the parking lot you think about how you really don't like to take painkillers. Even though you trust your doctor, you tell your car machine to look up back strain and the prescribed painkiller, to see if this is common. You are driving in a high-speed network region, so your machine performs a speedy search along the red links for "back muscles" and "painkillers." Seconds later it reports: "Yes, this painkiller is commonly prescribed for back strain for people of your age." It then gives a rating indicating just how common this is. You sigh to yourself in resignation. The Guardian Angel vibrates. You turn at the next light into the shopping center where the pharmacy is located.

Since you want to hurry home, you head toward the pharmacy's drive-up window. You swipe your insurance card through a scanner, which verifies your ID. The assistant pharmacist hands you a bag out the window; you thank her, and drive off. The transaction will be automatically sent to your insurer, and the $10 copay will be automatically deducted from your bank account.

Two days later, you get a message at home from your doctor's office. It has all the detailed results of the examinations, including images, in an attached hyperfile. Your doctor confirms that what you had was most probably diverticulitis. She advises you to read up on the right diet for this condition, and you click on the link she has inserted. It leads you to her clinic's page on special diets. You're glad that as part of the automatic exchange between the clinic's machines and yours, the hyperfile that includes this diet will be automatically transferred to your Guardian Angel. You'll be traveling next week for business, and you might forget what food to avoid at the various restaurants you'll be in.

Your handheld unit, with the Guardian Angel file in it, will provide an easy reminder. In her message to you, your doctor also asks if you want to be alerted to any new developments in diverticulitis diets and therapies. You respond "Yes." Her program creates an automated procedure that will monitor Medline on your behalf, alerting you of new review articles as they emerge. Every time a new finding is published that is relevant to the treatment of diverticulitis, the procedure will also send you the e-mail addresses of the experts who specialize in this treatment. This is such a common type of "alert" request that the clinic has routinely automated its e-form requesting the service on behalf of its patients. That's a real help, you think to yourself, as you stare back out at the still unfinished backyard.

Commerce

At the beginning of the 21st century, e-commerce was dominating all economic activity on the Internet. Start-up companies were being spawned at an alarming rate, so fast that you could no longer discern what business they were in. They became a blurred rush to a new pie in the sky. And what a big pie it was. The internal rate of return for the best venture funds in the United States exceeded 250 percent a year. Our MIT Lab for Computer Science, which had led to some 50 start-up companies in its 35 years, joined the crazy dance and fostered new ones, the most famous of which was valued by the market, at its peak, at $30 billion, an incredible 2 million percent of gross revenue. This prompted one of our people to utter a memorable phrase about a colleague one morning: "Today, Joe called in rich!"

Seasoned companies, seeing all this frenetic big-dollar activity, jumped into the fray, lest they miss the boat. All companies, big or small, mature or nascent, were after creating new markets. And hopeful investors were happily repeating the mantras of the spin doctors about a "new economy," where, just like in those unsolicited proposals I get about perpetual motion schemes, companies would no longer be valued according to fundamental principles, like the money they make, but by new magical rules built on wishes! From a short-term perspective, this bubble seemed to have no place to go other than to burst. From the longer-term perspective of preparing for a $5 trillion economy that will be waged over the Information Market-place, the value placed on some of the most promising of these companies did not seem so crazy . . . even without the new-economy justification.

Commerce is a huge arena, nearly twice as big as health care, if retail and wholesale trade are counted together. Let's see how a small corner of consumer and business dealings might fare in the new world of human-centric computing.

You and your spouse have decided to get out of North San Francisco and move closer to your jobs at the dot.com companies in rejuvenated Silicon Valley. You want to buy a house in the Palo Alto area. It's Saturday morning. You ask your personal information system, "What have you got on real estate purchasing in Palo Alto, California?" Your system, follow-ing its preset procedures, finds nothing in your personal Haystacks, and so visits the information systems of your friends and associates. It quickly spots a huge semantic link labeled "real estate" in one of your friend's records, and delivers the top-level views of the many linked threads under it. Among them, you find a Web-based service called Real Estate Associ-ation of Silicon Valley that describes houses on the market, and a private company claiming the same purpose. The latter attracts you because, you reason, it will have more offerings than the association, which is limited to licensed real estate agents. These days, lots of people are selling their homes privately.

You visit the company's site and find that it charges a fee to show you candidate homes, but it is low. You know that many such services are free, but you like what you see—the company does indeed list private offers and boasts a sizable record of successfully completed transactions, with praise letters from many satisfied customers. No wonder they charge something to discourage the idly curious. Your privacy monitor program, which handshakes with every site you visit, is giving you the green light, meaning that the service's privacy practices match your privacy constraints. So you commit, by saying "Go ahead, buy this." Your automated form-filling procedure completes all the service's e-form fields it can, with your name, address, and other repetitive boilerplate stuff, and then turns over the e-form dialogue to you. Fortunately, it is not complex, and you can finish it by speaking. You pick a rough price range, five locations you like, the size house you want, and a few more odds and ends about style and view. A minute later, you are browsing through three dozen one-page quad-charts. Each quad-chart features a picture of the house, vital statistics on its size and location, price and days on the market, and a verbal description. You are pleased to observe that there are quite a few houses that were placed on the market by their owners, without a real estate agent in the loop. You are also pleased that the quad-charts provide information in a consistent way, making it easy to compare features; evidently, the company asks sellers to meet certain uniformity standards, and the Semantic Web of the company's system captures and streamlines the sales notices of sellers that come through other brokers, converting automatically the format they use to a quad-chart.

You quickly narrow the search to 14 houses that seem promising. You and your spouse review the two-minute videos provided for all but one of them. A half hour later you have narrowed the search to 5 houses you'd like to pursue. You say so to your service and you hear immediately that you may proceed with 3 of them that involve private sellers, but should wait to hear from the agents representing the other 2, who have been notified of your interest. Within two minutes you are ringing the bell of the first house . . . without leaving your home. The owners are in and respond. You see and hear them and they see and hear you. After some niceties, you begin asking questions and taking in everything they tell you. As you talk, you ask them to show you different rooms and features of interest to you in each room, which they do with their roving video camera.

By evening one of the agents is prepared to show you the house she represents. This becomes a three-way collab affair with her, you, and the owners. The routine is similar to what you experienced earlier with the private sellers, except that for some questions the agent intervenes and *answers. By the following morning you have canvassed the five potential houses by machine and have settled on three that you would like to visit in person this afternoon. Just then, the real estate service sends the particulars of another house that just came on the market and meets your specs. Your service's automation procedure caught it. You take a quick peek at the house, but exclude it because deep down in your gut, you don't like it.*

You devote the rest of the day to the real physical visits, an hour's drive away. One house captivates you and your spouse. You really want it. It is privately offered, so you can negotiate right then and there with the owners. But you are not sure that the price you seem to be settling on is within range of other comparable houses. You pull out your handheld unit and ask the real estate service for a list of similar houses that sold recently. The results show the prices, tax bills, and locations. After reviewing them, you decide that you are within range. You agree on a price. Then you and the seller sit for a few minutes at his desk in the house's loft office (one of the features you love about the place). You contact your bank account and enter a password, then write an electronic check to bind the agreement. Your bank, using digital signatures, verifies the funds and sends the proper credit to the service's escrow account.

Still excited by the prospect of acquiring this dream house, on your drive home your spouse uses the handheld unit to notify the real estate service of the agreed-upon price. Its computer immediately asks if you would entertain mortgage offers from a few sources the company works with. You say yes, and within seconds you are staring at three offers, two subject to credit approval and one with a higher interest rate that requires no review.

Back home, you ask your living room computer to find out what your monthly payments would be for the different offers. You check your local bank online and find that the loan officers there are offering essentially the same deal. You decide to go with them since you have known them for a long time.

Your service is now contacting you to see if you want a list of surveyors and lawyers who can finish off the transaction. You select one of each, based largely on their customer ratings, which the service provides.

Six days later the house is yours. It was possible to close the deal fast because your lawyer's search for liens and other problems was nearly instantaneous, and he was able to obtain the necessary documents from the registry almost as rapidly. It was the surveyor who took four days to visit the place, but he then issued his report as a collab hyperfile full of voice and video fragments. He praised most aspects of the house, but gave a thumbs down to the basement sump pump and the water heater. You forwarded the report to the seller, and that evening sat down to talk with him while you both looked through the report on your screens. The seller agreed to drop the price to accommodate the repairs, entered the change on the contract, signed it digitally, and e-mailed it to you for your own electronic initials. You forwarded the final contract to the attorney. The closing took place the following day, with all parties in their offices and homes—a mere week after you started looking.

Just as human-centric computing sped the buying of a house, it will improve the efficiency of all sorts of business operations, including those that involve physical work.

Joe has worked for a major package carrier for 10 years. He is ambitious and intent on finding a spectacular way to help the company cut costs, so he can make his mark. He focuses on the company's delivery vans, where even the smallest savings per van becomes big when multiplied by the number of vehicles—150,000. He reviews in his mind how the vans are used. In the hours right after midnight, they are loaded with packages at numerous stations, for the morning's deliveries. The loading at each station is done by a crew of experienced, trained loaders who know the routes of "their" respective vans. They stack the packages on the van shelves in the right delivery order to speed the job of the drivers, who are already pressed to complete the entire delivery loop in time to begin the afternoon pickups. This requires the loaders to read the address on every package before deciding where to put it on the van shelves, and then to figure how to orient the variously sized packages so they'll fit the shelves and still be in decent order. As a result, loading is relatively slow— about 100 packages per hour by each worker. Joe's blood pressure sud-

denly soars—he remembers an experiment, in which unskilled workers were able to load up to 900 packages an hour without reading labels, analyzing addresses, and mentally figuring out how to position a vanful of odd packages. What if he could eliminate these steps?

Joe fires off a quick e-mail to Michael, a friend at the same level in the company, with whom he has drank many beers: "Do the bar codes on the packages tell the dimensions of each package?" The "No" answer comes swiftly and depresses him greatly. But he is not about to give up. He connects with Michael for a quick collab session and explains his idea, drawing sketches on their shared whiteboard, and asks him if he can think of any other way a machine could glean the dimensions of a package from available information. Michael says he can't . . . when a lightbulb turns on in his head. The bar code lists the weight of the package, which could be used to estimate its rough size, based on typical package densities.

Joe completes his original thought: If we could get the rough size of each package, then a computer simulation could be used to "load" the group of packages into a virtual van inside the computer's memory. In this simulation, packages would be positioned and repositioned so that after the whole van was loaded, they would end up on the right shelves and in the right order for delivery. Michael, excited beyond description, pipes in: "And then a voice synthesizer will speak the position of each package as that package arrives off the conveyor. If the loader hears the instruction 'shelf two, position four,' he will put the package where he is told, without having to read the label or figure out how to arrange the packages. Wow!"

Joe and Michael look at each other through their screens in stunned silence. Joe calculates in his mind that if they are right, the savings per van would be huge because the loading rate would be much faster and the loaders could be at a lower skill level—meaning, lower pay. This is too big for the two of them. They decide to bring in their friend Mary, who is an IT specialist at the company. She gets just as excited, and as the trio collaborates she initiates a search through her Haystack for simulations that involve package sizes. A few minutes later, while her friends *wait, she hits the jackpot: The company has exactly what they want—a simulation that infers rough package size from its weight, created two years earlier to estimate the volume of future package warehouses.*

Joe decides to cash in some chips he is owed by a loading supervisor. They bring him into the picture and tell him that they want to run an experiment. He agrees, and the following dawn, 10 local vans are loaded in a very special way. Their packages are all read by makeshift bar code readers before the expert human loaders place them in the vans. The readings are fed to the simulator program, which was modified by Mary for this new task. The results of the 10 loading simulations are now visible to Joe and Michael on the loading dock. They go into each van after it has been loaded, to see how close the simulation came to the actual way the vehicles were loaded by the specialists. This delays the vans from leaving on time, but that's part of the deal with the supervisor. The results are not optimal, but exciting nevertheless. The partners confer some more over their handheld units with Mary, now back in her office, where she uses the actual loading data to refine the simulators. After a week of successive trials, the software is in much better shape. Even the loading supervisor is surprised that the simulation comes so close to the way his loaders fill the cars.

All four people go to the operations vice president with their results. He can't believe his eyes, and quickly decides to swing some serious engineering resources in this most promising direction. Four months later, special bar code readers, loading simulators, and speech synthesizers are installed at 300 trial sites. The system proves itself and moves to full deployment in another eight months.

This "little hack," as we would call it at MIT, is a variant of a true story. It increased the actual loading rate from 100 to 400 packages per hour. The savings, multiplied by 150,000 vans and compounded by a 30 percent reduction in labor rates, resulted in an estimated annual savings to the company of $400 million per year.

Joe and his friends have made their mark. Their ingenuity was the principal factor in their success. The human-centric systems they used were in the background, where they should be, helping them easily try different approaches and work with one another, without the machines getting in their way.

To get a sense of how widely the human-centric forces can help us do more by doing less, let's take a peek at a few more applications, in abbreviated form.

Disaster Control

A major earthquake strikes a West Coast city in the middle of the night. Automated emergency procedures in various municipalities are triggered by interconnected seismometers. The procedures were developed independently in the human-centered computer systems of local fire stations, hospitals, police precincts, the Coast Guard, the National Guard, and emergency medical teams that must now rise and work together to combat this natural disaster. The city's disaster control center starts receiving information automatically from all these sources and many other interconnected physical devices, including wireless, battery-powered cameras and microphones, that provide data about rapidly changing conditions. The metadata tags provided by these weather sensors, traffic flow meters, emergency room registration analyzers, and 911 call logs are invaluable in helping to sort the massive information as it arrives. The flow is overwhelming. The dozen employees who are responsible for operations can barely keep up, querying the vast amount of data, even as it gets threaded by automatically generated and manually provided red links. The mayor, the governor, and police chiefs are on their portable units and join in secure collab sessions with the people at the disaster control center. Together they monitor local situations by viewing images and assessing the data that clusters around each major problem area. They watch and digest simulation results that predict the changing needs for backup power, temporary shelters, crowd control, and traffic redistribution. In some cases, before firing up a new operation, they quickly scan the procedures and outcomes of prior disasters, accessible through Seman-

tic Web links. They issue warnings through radio and TV stations, even as they call for additional resources from the federal government.

This example of free exchange of information is based on computer research scenarios for handling disaster control.

Medicine in the Bush

Modumba is constantly coughing. He is quite sick. His brother carts him to the medical shack in the African village center at 6 A.M. and wakes up Rapilla, the man in charge. He is not a doctor, but he has been taught how to use the local X-ray kiosk, a gift from a Belgian hospital, donated to take care of such emergencies. The X ray is ready in a few minutes. Rapilla places it against the window, lifts up an inexpensive digital camera, and takes a couple of pictures against the dawn light, which are fed into the kiosk. He then speaks to the machine, asking for a connection with Dr. Hamish Fraser in Boston. He is grateful that he can speak and doesn't have to type, for he can neither read nor write. Dr. Fraser, as he gets up from his living room chair, muses that he wouldn't have to do so if he hadn't pioneered this $500 X-ray transfer scheme, to eliminate the cost of the $30,000 professional X-ray digitizer. But he is anxious to see what this new case has brought. He looks at the photos from the African village he knows well and spots a suspicious shadow. He asks Rapilla to center the digital camera one hand to the right of the last shot, and increase the camera setting to full zoom. A minute later, Dr. Fraser's suspicions are confirmed. There is a clear evidence of tuberculosis. He tells Rapilla that the patient will need a six-month course of antibiotics, and will have to go to the nearest town hospital for more extensive tests and treatment, even though it is a day's journey.

This scenario, minus the speech aspect, is based on actual work of Dr. Hamish Fraser, a medical doctor and computer scientist at LCS.

Total Financial Services

Silvio and Mary Berini have been dealing with the Central Financial Group (CFG) ever since they were in their teens. They have a warm feeling for their account manager, Wilbur, whom they have met in person several times. But they also feel friendly toward the rest of this great company that, over the years, has become familiar with their financial needs and problems, and has consistently helped them with practical advice, loans, insurance plans, inheritance taxes, income taxes, banking services when they travel, and a whole lot more— "everything that has to do with money," as CFG's motto says. They are now speaking to their system, where the CFG application is running, because they have accumulated some extra savings that they would like to invest. They are asking questions, some of which are answered by the local application, while others are passed to and answered by the CFG corporate machines. The combined systems know a lot about the Berinis, and after comparing financial metadata, offer four possible investment categories that fit their current financial situation, goals, and risk tolerance. Silvio and Mary drill down into these investments and pose questions using CFG's Semantic Web. At one point, CFG can't provide the answer and forwards them to the machines of the company that interests them, where they get all the information they wanted and a quick online chat with a customer representative. They are finally ready and call for Wilbur, who comes promptly to their screen, even though he is across town from his office, driving his car. They discuss with him their quandary between two equally attractive options—one a stock, another an aggressive money market account. He quickly settles their question, and concludes the transaction. A bunch of automated procedures are immediately generated by the CFG software to monitor the stock the Berinis just bought, and other competing stocks that may do better, and alert them, and Wilbur, to any significant changes. Silvio and Mary are thrilled that they get a seamless, highly customized service from what they know are infinite mountains of information along many bulky core services, like life insurance, medical insurance, retail banking, brokerage, and so on, which constitute the inner strength of CFG.

This imaginary scenario was concocted out of the long-range plans of two financial institutions and an insurance company.

Play

It's Saturday morning. The 24 teenage boys cleanly divided into two rival camps turn on their machines, each in his home. Most of their parents sigh variations of the ancient lament: Why do they have to play these fierce testosterone-laden games? Why can't they do something healthy outside? The saving grace is that their children are members of computer, rather than street, gangs. The machines are on. The boys, who will each fly futuristic fighter craft, flip down the visors of their helmets, which show the black deep-space battleground in 3-D virtual reality, and communicate wirelessly with their powerful basement computers. One boy on Team Warrior who is at his grandparents' house for a family celebration later that day, is limited to the old-fashioned desktop screen, but at least he is able to download his spacecraft profile off the Net and link into the game through the nomadic fighter pilot program, so his team wouldn't be a man short. The rival captains give the starting signal and the battle begins. On everybody's visors realistically rendered spacecraft appear, with guns ablaze as each player tries to down enemy craft while hiding behind space debris, or executing unpredictable maneuvers with the thrusters. The cameras on the wall screens with which the helmets communicate pick up the facial expressions of the players, which are eminently discernible, despite the visors, and show them to their adversaries and their partners for heightened intensity. Little collab groups spring up, on the fly, as members of one gang conspire for a maneuver against the enemy, which they had practiced the previous day. The maneuver is successful. Only seven spacecraft are left. Each pilot speaks commands to his craft and pokes at radar and other info that also appears on his visor, to gain every possible advantage on his adversaries. Three boys engage in a bit of disinformation and trick their rivals into an open area where they shoot them all down. Victory for Team Warrior. The boys smile, take a deep breath, and get ready for a new game. And they are happy—has anyone earned the right to moralize upon their shenanigans?

This free-information-exchange distributed game is an imaginary scenario based on plans reported by video game makers.

Sundials

The first-year college students arrive at their "Hanging with the Hackers" class. They love this class because of its focus on creativity and total openness. Two weeks ago they had to design and build the best musical instrument they could with materials costing less than a dollar. Colette won, turning a fat straw into a flute with only a few clever incisions. Last week they had to design a round restaurant table that could change diameter smoothly from two to three meters and cost under $500. Today, the assignment is to design a sundial and build it out of wood and nails; the students' creations will be tested on the windowsill at the end of the three-hour class, by three teacher judges. The students group into threes and begin scrambling through the school's Haystacks and over the Web to learn how the sun moves. Who would have thought that it doesn't move smoothly, but hesitates and speeds up depending on the season, sometimes by as much as 15 minutes. And why are there so many different sundials—polar, equatorial, vertical, horizontal? The kids know that they must make choices and decisions. They keep probing via red links the literature on sundials, they ask questions of their machines to initiate searches and test assertions, and try designs on the simulator provided by the professor in charge. But their most important activity is collaboration—amongst the partners in their group and with other kids in a few allied colleges that are, by plan, working on the same project at the same time. The students are learning fiercely, not only about the subject, *but also about teamwork, without realizing that they are doing so. Before they know it, time is up. The sundials must go on the windowsill aligned with the north-south line for the contest. The judges read the times they show. Once again it is Colette, this time with her group, who wins—their dial is only 2.5 minutes off. The classmates grouse, "Why is she so 'lucky' so often?"*

Except for the use of human-centered machines, the events in this story are true and arose in the MIT freshman seminar I occasionally teach, "Hanging with the Hackers."

Why These Five Forces?

I could go on and on with examples like these. Human-centric computing helps us do more, in natural, almost invisible, ways, to advance whatever activity we might be pursuing. That's to be expected—and the reason we have gone after human-centric computing in the first place.

One question that arises, though, is why there are five human-centric forces. Is there anything special about speech understanding, automation, individualized information access, collaboration, and customization? Or are these the first ones that came to my mind?

There is a sensible reason. Imagine tomorrow's Information Marketplace, where a billion or so people and several billion machines are interconnected. As various people and machines interact, they will fall into only one of three categories: human-to-machine, machine-to-machine, or human-to-human. And when more than two entities interact, they'll still involve these basic pairings.

The five forces sustain the principal ways these categories become useful to people. Human-to-machine interactions beg that people communicate with their machines in the most natural way possible, which for us is talking and viewing. The human-machine category also includes finding information from the vast store of data in our machines and on the Web; that's where individualized information access comes in. The same category calls for adapting programs and data to people's individual needs—that is, customization. The machine-to-machine category involves only machines working with one another to do our work—that's automation. And the human-to-human category involves people working with other people, which is supported by collaboration.

The five human-centric forces are fundamental because they cover the only possible categories of interactions among humans and machines. Even if additional forces emerge, as they might, they will serve one of these fundamental categories. Meanwhile, the coverage of human needs, made possible by the five forces, should be enough to help us explore human-centric computing and even achieve a good part of its promise.

Dovetailing People with the Forces

Yet, even in powerful combinations, the human-centric forces aren't quite as strong as they could be. In all of the preceding scenarios, the resultant utility was due not just to the new technologies, but also to the way they were dovetailed with human procedures—the doctor's one-on-one exam, the face-to-face negotiating for a house, the human loading of a van, the police coordination of a response to disaster, students learning as a group.

I now beg to be excused for distorting beyond recognition an Asian proverb that originally dealt with the need to understand oneself before venturing to change the world:

> *When our ancestors hit upon leather technology, many people proposed we cover the countryside with leather, to keep the stones from hurting people's feet. Others argued, it would be better if everyone wore leather shoes.*

After a few millennia, we've ended up doing both—wearing shoes *and* paving our sidewalks! This is what will happen between tomorrow's human-centric technologies and the people who use them. Computer applications will rise closer to people through all that we have discussed so far. And the people who use these tools will do a little adapting of their own to meet the machines' capabilities. Think of this process as tuning and dovetailing human and machine procedures to get results that really hum.

But wait a minute! Isn't this heresy, coming after six chapters of advocating that machines should serve people rather than the other way around? Not really, because left to our own devices, we'd cover the countryside with leather—we'd blanket our world with computers, without making any human effort in their direction. But that won't work, because machines are neither as intelligent nor as flexible as we are. So we must step down a bit toward them—just a bit—as they undertake their long march to approach us.

There are bigger reasons for dovetailing people and machines. In an organization, people work together to achieve goals they could not reach by themselves. The interactions among people and machines that lead toward shared organizational goals will be pivotal to the way the organization performs. Fine-tuning this interaction is the ultimate practical action we can take to do more by doing less. A computer system, however new and capable, is ultimately available to everybody. The winners among otherwise equal competitors will be the ones who can achieve the best dovetailing among their most precious asset—their people—and the new human-centric technologies.

Let's take a simple example from today's Web. I have already mentioned Bibliofind—a collection of several hundred antiquarian booksellers united under a common search engine. The ensemble is very useful in locating a book you want. However, past that point, the one-organization illusion breaks down. The service provides a shopping basket in which you can accumulate your purchases. Say you order 15 books and enter your credit card number. But a day after the magical click that submits your order, you get several e-mail messages from different bookstores telling you that the credit card you used is unacceptable. It turns out some of the booksellers accept credit cards, while others don't want anything to do with them. What started as an enjoyable, convenient experience has suddenly become a nightmare of work; you must go back and dredge up all the individual sellers' addresses, determine who wouldn't take credit cards, and mail individual checks to each one. This is clearly a situation where the human and machine procedures are badly dovetailed. If another service were to come around with a common payment scheme it would win hands down.

The story I told earlier about my exhausting, late-night hotel reservation fiasco is another example of badly dovetailed human-machine resources. The check-in clerks do a lot more than they should. They enter credit card numbers separately from reservation information already given by the arriving guest. They "print" magnetic card keys and envelopes for keys, manually. They communicate with housekeeping via telephone rather than relying on a well-tuned procedure like a maid pushing a phone button when a room is ready.

Airlines offer perhaps the most scandalous examples of mismatched human and machine procedures that result in long, unbearable exchanges at the counter. It's almost as if airline executives go out of their way to maximize the waiting time of their customers, by throwing piles of software together and leaving the searching through all the layers of this complex miasma to the overworked clerks. "Your fare, sir, will be $2,300." "Thanks, but do you have a D or K fare?" "Let me see." Sixty keystrokes later: "Yes, indeed, I can offer you the special discount fare available for this period at $720." In this, the dawn of the 21st century, why can't the airline software search for the cheapest fare? Millions of people throughout the world are spending a half hour or more every day waiting in long queues at hundreds of airline terminals, when with the right human-centric technologies they should be able to complete the entire ticket purchasing or check-in process in 30 seconds! We should favor with our business the airlines that wake up and go human-centric, reducing our waiting time, and serving our purposes.

Bad dovetailing is also what leads to those maddening automated answering services that tell us to push all the buttons and wait, as well as the ridiculous instruction from the washing machine repair service, or the furniture delivery service, that you should plan to be home for them to arrive "any time between 9 A.M. and 5 P.M." Sure, just kill an entire day. As much as services have advanced in the industrial world, they are still well behind what could be done to truly serve the needs of the customer, instead of cagily serving the provider's own needs.

The service dysfunction that we are observing is, perhaps, understandable at a more philosophical level. Consider the United States of America, which already embodies what is happening throughout a world that is becoming increasingly democratized and wealthier. The principal credo of U.S. citizens is individual freedom—something about which they are prepared to fight with their lives if it became necessary. How can a person or group whose central belief is freedom be simultaneously a top-notch servant? I don't think it's tenable, even if the provision of good services can lead to greater profits. People, especially as they get wealthier, will not want to "compromise" their core beliefs for anything, even monetary gain. Here then is another good reason to adopt human-centered systems: Their growing servitude to us will counterbalance the declining quantity and quality of services people will be willing to offer to their fellow human beings.

The desire to better dovetail human and machine procedures might suggest the need for certain centralized tuning practices, reminiscent of the time-study procedures people imposed in industry to make sure workers were utilized at their optimal level. While such approaches may work sometimes, they won't in most cases. There is seldom a consultant, or a CEO, who knows better than the workers of an organization how to improve the performance of their individual tasks, or even of the entire organization.

In organizations, the management secret to good human-machine dovetailing lies in letting the employees largely determine what system resources they need to make their jobs easier and more productive, maximizing the performance of the overall enterprise. That doesn't mean that a "tyranny from below" should rule all decisions, since some centralized coordination will be inevitable. Rather, it means that the initiative for utilizing information resources should be distributed, with senior management having its ample share. The earlier examples of the shipping clerk who ordered some custom hardware and software to speed work and reduce cost, and the package delivery employees who saved $400 million for their company, make this point. Imagine all the people of your organization bent on acting in a similar way to improve their respective jobs and the organization's business. The human-machine tuning that would emerge from such decisions would be natural and would parallel Adam Smith's "invisible hand"—a magical, self-regulating mechanism that would help dovetail people and machines to one another without a central directive, but under shared organizational goals.

It's never too early to start dovetailing people and machines to advance your organization. If you are a manager, you can begin, even before you acquire a new breed of systems, by giving your people the power to select what they think is best for their individual activities. Then form task groups that will try to coordinate among the different uses of information in your organization, to maximize the dovetailing.

Don't expect your "information technology" people to do this by themselves. They can't, for they do not know the organization as well as the employees who run all the parts of it. You may use the IT people as resources, but not as drivers of the changes that you need. An approach I outlined earlier works here equally well: Form teams among people who are youthful in attitude, who know their organization well, and who work easily with others. They need not be technical, though some knowledge of IT among them is useful. Their task is to look at the ways human and information resources are or could be intertwined throughout the organization, toward improving the organization's goals.

The dovetailing of human and machine resources is what will make the difference between winners and losers in the Information Age. Everyone will have essentially the same tools. Those who will shine are the ones who will blend the new technologies with themselves.

We have talked enough about human-centric systems. It's time to go build them.

Eight
OXYGEN

How do we transform the vision of human-centric computing into reality? There is only one way: Build and test a prototype. That's how all computer systems have evolved. A project called Multics developed the concepts used in the popular Unix operating system. Project Alto led to the Macintosh operating system, which in turn led to Windows. And the fledgling Arpanet begot the Internet and the Web. Each of these grand projects involved hundreds to thousands of person-years of work. Each process was also fickle; because computing systems are highly complex software mechanisms involving millions of "moving parts" (lines of code). Their behavior cannot be accurately predicted until they are built and tried. This is different from industrial-age products like cars and watches, which, relative to software, have fewer parts—in the few thousands—and can be designed with much greater confidence to provide a predictable level of performance.

Designers face additional complications that pop up in the course of the software development process, as decisions are made to avoid difficulties or to pursue opportunities, none of which were foreseen at the beginning. A good deal of the research and development associated with a brand-new software system involves managing this terrain of unplanned consequences, with the result that new capabilities are added and expected capabilities are eliminated. As people use the prototype they find problems and suggest changes that modify the plan even further. Indeed, the development and early use of real systems are inexorably intertwined. Neither can proceed without the other. They both move forward through a succession of iterations. By the time the system has begun maturing, it is likely to differ significantly from its original concept. Sometimes this unpredictability leads to failure. At other times it gives rise to great success.

Multics was built in the late 1960s at MIT's Project MAC—the precursor of today's Laboratory for Computer Science (LCS). It was originally intended to create a way for a few dozen users to share a very expensive processor. The work introduced many innovations to the world, with names like file directories, access control, and virtual memory. But the approach didn't take off commercially until it was simplified and radically changed into Unix by AT&T Bell Labs, a collaborator on the project. Even its new name was a pun designed to reflect the simplification that its designers introduced—to their ears, "Un-ix" sounded like a merciful single thing, contrasted to "Mult-ics," which was a multitude of things . . . whatever they may have been.

Alto, pioneered by a group of researchers at Xerox PARC in the 1970s, explored easier ways in which people might use machines. It introduced to the public the familiar ideas of menus, windows, and pointing by a mouse. However, it, too, didn't succeed commercially until Apple Computer transferred the key ideas to the Macintosh personal computer. Later, Microsoft Corporation made further changes to arrive at its Windows system.

The Arpanet began in the late 1960s under the Defense Department's Advanced Research Projects Agency (DARPA). It was built by Bolt Beranek & Newman to interconnect large computers, mostly at research institutions, so the researchers could share each others' machines instead of constantly asking the government for money to buy new machines. But only after the Arpanet was transformed into the Internet, a system to interconnect computer networks (instead of individual computers), and after the Ethernet appeared and fueled the growth of thousands of local networks, did the technology become widely spread. And only after the arrival of the World Wide Web in the early 1990s did the Internet-Web pair become widely known.

Prototypes open doors with new ideas. Success, if it is to come, follows many additional transformations and a good deal of serendipity. That's what I expect will happen with the evolution of human-centered systems.

The human-centric approach has been taking shape in my head for more than a decade, although I didn't hit upon that label until very recently. One of my earliest forays was at the 1992 LCS annual meeting. Some 50 leaders of LCS research and their families had descended on Cape Cod for three days of presenting ideas, discussing strategies, and playing games on the sunny beach and ocean, as we do each summer. I was nervous because I had finally summoned the courage to advance before my peers the idea of searching for a radically new computer language and operating system that worked at a level much closer to that of the user. I was laughed at. My technologist colleagues could barely contain their indignation, especially the ones who were systems specialists. After years of catering to difficult and complicated problems concerning file systems, security, communications protocols, applications interfaces, and the other technical aspects of the software that surrounds computers, they were not about to buy into what they perceived as unadulterated bull—some mushy phrases by the lab director about ease of use and increased human productivity.

At one level, I was sad they didn't see the future as I did. But I also found hope, because I smelled a "Schopenhauer." The great philosopher observed that every important new idea, when it first appears, is ridiculed. Yet after a while, everyone agrees that it was all along an obviously great idea. Of course, not every idea that is put down turns out to be great. I have proposed and heard quite a few ideas over the last four decades that were rightfully ridiculed. Besides, at LCS, starting a statement toward a colleague with "No, no, no! You don't understand," and then explaining why, is our own sweet way of mercilessly probing the strengths and weaknesses of a new idea—an essential process for filtering out the promising from the nonsensical. None of us views these attacks as personal. So I was bothered but not distressed by my colleagues' reaction.

That meeting had followed a 1991–1992 study for ARPA, which I had instigated and led, on how to make computers dramatically easier to use. It led to the notion of "gentle-slope systems." This new breed of system would give instant gratification with every little advance in learning that a user made. And each additional small step would allow a user to achieve even more by expending relatively little added effort. Spreadsheets and some photograph manipulation programs have this gentle-slope property. A small group of believers formed, but the disbelief from outsiders persisted. It seemed that technologists in the early 1990s were passionately against tackling the ease-of-use issue, because they did not perceive it as a problem amenable to technical solution.

This resistance was at odds with what my gut told me—which I had gradually come to trust more than the prevailing wisdom. With computers becoming more complex, and with technology advancing rapidly, if we kept going along the present path we would end up with ever-greater problems in the use of machines. I was becoming increasingly convinced that the advances in technology, instead of being laid raw upon the users, could and should be steered toward harnessing machines to serve what people wanted to do. So I kept putting up at each annual meeting of our lab the strategic wish that we develop approaches that make machines easier to use.

By 1995–1996 the negative reactions surrounding these ideas started giving way to positive interest. In Davos, Switzerland, at the World Economic Forum, a group of CEOs in the computer business concluded that the biggest issue facing them was making machines easier to use. At about the same time, as I was finishing *What Will Be,* a book on how the new world of information would change our lives, I was struck by a surprising observation. A handful of new approaches seemed to lie at the heart of every new application of information technology that I could envision for the future—whether in commerce, health care, recreation, manufacturing, education, government, or any other service. They were speech understanding, automation, collaborative work, and customization—four of the five forces I now call human-centric. I pulled them together and began asking businesspeople to pay attention to them in their practices. A glimmer of a nuts-and-bolts approach toward bringing information technology closer to people had appeared. But it still had not distilled itself into an action plan.

The human-centric approach took its final shape in 1999 when DARPA's Information Technology Office, under the leadership of David Tennenhouse (originally an LCSer, who is now VP of research at Intel), issued a call for proposals on "New Expeditions." It asked for radically new ways of deploying and using information technology, unfettered by today's constraints but within the realm of possibility. This was a different tactic from the agency's normal mode of compartmentalizing research in different technical areas. It reopened DARPA's door to bold and sweeping new visions that could not have been proposed under the old approach. Excited by the prospect, I convened our top LCS leaders to put together our "expedition" proposal.

Such brainstorming sessions are a great deal of fun and alive with movement. Some ideas tossed into the hopper by different people generate excitement, some cause violent turns that leave other ideas behind. As we went through this and subsequent exercises, my colleagues Anant Agarwal, Rod Brooks (who joined us a bit later), Frans Kaashoek, and Victor Zue, who with me put together the vision we called Oxygen, began feeling the exhilaration of an approaching big change that could make a real difference in people's lives. Among technologists, this "high," which is felt at the threshold of something radically new, is similar to what artists feel as they create a new form of expression, or scientists before a fundamental discovery.

While we shared a broad overall view, to each of us, Oxygen became a placeholder for strong individual beliefs. Anant, who contributed the name "Oxygen," was fascinated by a world in which technology would be abundant and pervasive, like the air we breathe, and he wanted to build machines out of a new technology he pioneered. Rod, who heads the MIT Artificial Intelligence Lab, was focused on using vision and AI techniques to make the computer vanish from its "temple," where people had to go in order to use it. Frans was excited by the prospects of distributed networks and secure file systems, and Victor, steady on his course for more than a decade, was after speech as the dominant mode of interaction. I was stubbornly, some would say maniacally, after one thing—the human-centric focus, through the five user technologies discussed in this book. We argued a lot, each trying to convince the others of the centrality of our individual views—tensions that are essential to the healthy evolution of a major research project.

In the end, we agreed that our unifying goal was pervasive, human-centered computing, which we would tackle by building a prototype we would use in our daily lives. Thus Oxygen was born at MIT, as a five-year research project. We became so enthused that we vowed to build this prototype even if we could not fund it on the first try. We felt so passionate that in early 1999, we made Oxygen the centerpiece of our laboratory's 35th anniversary celebration, even before we, together with the MIT Artificial Intelligence Lab, submitted our proposal to DARPA. As things turned out, the proposal was approved and the project was formally started in September 1999. By May 2000, Acer, Delta Electronics, HP, Nokia, NTT, and Philips became our industrial partners, forming the Oxygen Alliance. The $50 million project, building up on the combined strength of some 250 researchers, was finally on its way.

Great ideas, enthusiasm, hard work, and money are essential for the evolution of a grand new project. But they do not alter the cold fact that Oxygen is a proof-of-concept test bed. It could fail just as easily as succeed. After all, if a project is safe, it is not worth tackling, for it usually makes incremental improvements on old ideas. And if it is too crazy, it is doomed to failure from the beginning. At LCS, we are after radical change with projects that have a chance of success in the one-in-three range, and a time horizon that spans 5 to 15 years. Oxygen fits this mold. I will describe it here briefly, to show how real hardware, software, and communications could be put together to achieve human-centric computing, and because it is the human-centric reality I know. Other labs, consortia, and companies will pursue competing research projects to achieve similar goals. For example, Carnegie-Mellon University, the University of California at Berkeley, the University of Washington, and Georgia Tech are involved in work aimed at making computers pervasive and ubiquitous in our world, with an associated intent to also make them more usable. Please keep in mind that even though I use MIT's Project Oxygen and my colleagues' work as examples, I do not speak for them or for the project.

How does a system like Oxygen get out of the lab and into the commercial world? Like Multics, Alto, and earlier systems, Oxygen may lead to corporate start-ups, or be picked up and made commercial by software companies. We will try to scare away failure by building both conservative and radical alternatives for each component technology. The conservative "safe bottoms," as we call them, ensure that the pieces of Oxygen work well individually, with modest goals, and therefore can be integrated to test the capabilities of the whole system. The "dangerous peaks" give us the high-risk, high-payoff opportunities to advance dramatically each of Oxygen's component technologies. Ultimately, however, avoiding failure is not a priority. We are driven above all else by the excitement of new discovery, especially before seemingly impossible tasks. And regardless of whether Oxygen succeeds or not as a practical system prototype, we expect that its key ideas will survive.

These ideas arise from the many researchers who work on Oxygen—the proud parents of the new system. Without these people, the project would not have been possible. For my part, I will be ecstatic if this effort inspires us all to launch a new era of human-centric computing that begins to finish the Information Revolution.

Putting It All Together

Oxygen combines under one cover the primary software technologies that bring machine capabilities closer to people, the hardware "delivery vehicles" that let people use this software, and the core software that pulls all the different pieces together.

Oxygen's hardware and software enable people to interact naturally through speech and vision; automate human actions; provide individualized access to the information they need; help them work with each other across space and time; and customize machines to their unique desires. The system also handles people's growing mobility and helps them control their physical devices.

Oxygen is a prototype, not a commercial-grade system like today's Unix, Mac OS, and Windows. More than a thousand person-years are normally required to give a software system that level of stability, and over a decade of real use is usually needed to iron out bugs and shape it into a mature form. Oxygen's principal mission is to demonstrate a radically new approach of using computer technologies to serving people's needs. This means that as the exploration proceeds, the system will change considerably from the early description that I am presenting here.

Compared to today's computer systems, Oxygen brings many changes. Instead of you going to the machine, as you always have, the system is now all around your human world, ready to handle your needs. Interactions between you and the system become natural through speech and vision, causing your mind-set to shift from cognitive to perceptual, or as the Philips people say, from "lean forward" to "lean back." You no longer have to plan what you'll type, you just react to what you see and hear by holding a dialogue with the machine. Devices, especially for mobile use, become anonymous and acquire the "info personality" of whoever is using them at the moment. Security, too, becomes person centered rather than device centered. The meaning of information becomes more important than its structure. Resources, like people, information, and machines, are located by intention. Computing resources become plentiful, closer to the notion of electrical sockets and further from the notion of a digital shrine to which we make pilgrimage. Software becomes pervasive, embedded in physical devices everywhere. It also becomes nomadic, following you around, updated on the fly as needed. And it becomes "eternal," running "forever"—always there, always "on," with no need to ever restart any machine. Most important, you and your application programs use Oxygen through its human-centric technologies—the "gas pedal," "steering wheel," and "brake"—that bring the computer up to your human level, to serve your needs.

The Handy 21

Oxygen has three key delivery vehicles. The first is the Handy 21 (which derives its name from being handheld and from the 21st century). It is a powerful, portable device of the same size as a cellular phone, which you would carry in your pocket or purse. Its purpose is to provide you with whatever computational and communication resources you need when you are away from your office, home, or car.

The battery-powered Handy 21 consists of a microphone and speaker; a small screen on which you see text and pictures; a miniature camera to which you can show things and people (including yourself); antennas for communicating with different wireless networks; and a few other odds and ends. It has no keypad of any kind. You communicate with the H21 through spoken dialog and by viewing what it shows you.

Like a chameleon, this little gadget can change function under the influence of software that flows into it. It can be a high-speed network node when you are in your office building or home, communicating rapidly with computers at these sites, or a somewhat slower network node when you are outside but still near these principal resources. When the H21 detects no such computer network around, it "sniffs" the air for the next available communications medium—usually a cellular phone network—and changes itself to a cell phone capable of communicating in that system's protocol, be it European, American, or other. The H21 can also become a two-way radio for talking with other nearby H21s, perhaps in the Sahara. It can even turn itself into an AM or FM radio, or television, with the right software. The H21 makes all these changes invisibly to its user, under the control of system software, based on what's available and what's best for the task at hand.

The H21 takes a step beyond the various handheld, fixed-function devices of today—the high-powered cell phones that can also access the Web, or the palm organizers that can be used as browsers and handle e-mail, and even the wristwatches that double as cell phones. The H21 can implement any and all of these functions, and many more, with the right software. We want the Handy 21 to have this huge flexibility because it is centered on a person—you—and there is no telling where your mobility may take you. The H21 has to be prepared for every communication eventuality it may encounter.

But how is such a small device expected to do all these things? It will change function using a new approach, called SpectrumWare, pioneered by David Tennenhouse and John Guttag of LCS. Except for about 10 percent of the H21 hardware that is analog (its antenna and so-called frequency conversion and analog-to-digital circuits near the antenna), the rest of the "circuitry" is all digital and programmable. To change from one function to another, the H21 simply loads new software into its digital circuits. Some of the software that gives the H21 its most essential functions is stored within it. But a lot of software will flow into it, nomadically, for less frequently used functions.

When you are on the go, this portable unit is a single point of contact between you and the world of information. We have chosen this approach because we want to be all encompassing, as we learn what is and is not useful to people in practice. An alternative involves several small units you carry with you, which are interconnected by a low-power network and a personal router, perhaps woven into your clothes. Such a body net would let you go "distributed," picking up whatever hardware you may need. For example, if you are going on a long boat trip in the Antarctic, all you might need to take along is a small Global Positioning System receiver to pinpoint your whereabouts when you venture away from the mother ship, and the H21 software that turns your portable into a medium-speed network node so you can communicate with your ship's people and computers. Dave Clark of LCS, who is a champion of this approach, is pursuing such a distributed version of the H21.

For now, the Handy 21 is being built out of COTS—"commercial off-the-shelf components." Which means that it will be large and heavy . . . and a safe bet for contributing to the overall system. But consistent with the Oxygen philosophy, there must also be a riskier, higher-payoff challenge. In this case, the challenge lies in developing a very light unit that consumes much less battery power, yet is fast and powerful, regardless of the function the H21 is mimicking. We hope this change will come about through the use of a new breed of computer chips, called Raw, pioneered by Anant Agarwal and Saman Amarasinghe of LCS. Today's chips process signals the way city streets process cars. For a signal to get to the right place, it has to check at every intersection whether it should turn right or left to get through the chip's many internal "wires." By contrast, in a Raw chip, the software logically rearranges these internal wires, so each signal knows ahead of time all the turns it must make and can zip along without having to slow down.

Each software application can reconfigure the Raw chips in the H21 so they are optimized to carry out the calculations which that particular software needs.

So when you want to collaborate with a friend, and you call up a collab editor, before that program runs it will tailor the Raw chips' circuits to its needs. This optimizing can reduce the power and increase the performance of the H21 by an incredible 100 times . . . for that application only. When you choose a different software application, it will reconfigure the Raw chips accordingly, just before it runs. By exposing their "wiring" to the software, the Raw chips allow themselves to be customized to suit the needs of any application. The Raw chips are attractive for human-centric purposes because they give the H21 so much more performance at greatly reduced power. Our portables will last much longer on their batteries, and their wireless forays will reach greater distances at higher speeds.

The Handy 21 also employs sophisticated strategies to manage its wireless transmission by changing power, frequency, and data rate, and it trades off computation, communication, and power consumption to maximize its utility. What the H21 does in these situations is similar to what you might have to do if you want to be heard in a noisy environment, while conserving your strength. You might raise the pitch of your voice to stand out from the drone of predominantly deep male voices, or you may slow down your speaking rate, or you might do both. And instead of wasting your breath asking someone for today's date, you might "compute" it yourself by looking it up in your calendar. The H21 is constantly pulling tricks like this, which are an integral part of the Oxygen system software.

Another interesting property contributed by Dave Clark is the H21's ability to capture the identity of a physical device at which it is pointed. This can be done if an infrared, radio frequency, or visible bar-code "tag" is pasted on physical objects at which the H21 may be pointed. In a long corridor, you might point the H21 at an office door and see on your portable's screen the name of the office's occupant, complete with photo and title. Or you might point your H21 at a printer and say, "Print my last memo here." It's even possible to point an H21 at a washing machine to read the label on a malfunctioning part. The H21 would then contact the manufacturer over the Net for instructions on how to fix it. These could be displayed on the H21 screen, showing you which screw to turn to fix the problem. Being able to point at physical things is a natural, and hence easy, way to bring physical objects we care about under our control, in addition to using automation.

The Enviro 21

Unlike the Handy 21, which goes wherever you go, the Enviro 21 is a stationary computer in the walls of your office, in your home basement, and in your car trunk. It derives its name from being centered in your environment, rather than on your person.

The purpose of the Enviro 21 is to provide you with ample computational, communication, and perceptual resources in your normal work and living environments. The Enviro 21 has the same capabilities as the Handy 21, but packs much more punch. It has a massive capacity for storing information, much higher processing and communications speeds, and is connected to a wide range of powerful hardware accessories. You can think of the relationship of the E21 to the H21 as the relationship of a power outlet to a battery.

Like the H21, the Enviro 21 is initially built out of COTS components, and later Raw chips. It is connected to wireline and wireless networks and provides you with connectivity to the Web and the world's networks. Your E21s also provide support for your H21s, by storing your info personality and the nomadic software your portable may need while you are on the move. And when you are walking in your building, nearby E21s will off-load power-hungry computations from your H21.

An important property of the E21 is its ample "tentacles," which are connected via wires or wireless links to devices and appliances. This is the main way in which Oxygen interacts with the physical world around you, complementing and enhancing the H21's ability to point. In your office, these tentacles may reach to the phone and the fax machine, the electronic whiteboard, the large human-size display screens in the walls, all sorts of printers and scanners, as well as controls for the room temperature and humidity and even a sensor that detects the position of your office door; don't forget that when your door is open, it signals your local and distant coworkers of your willingness to be interrupted. In larger offices, the E21 will be distributed all over the room. Camera and microphone arrays in the walls will track individuals as they meet around the room, use a whiteboard, or point to different objects. Lip-synching of what a camera sees to what a microphone hears could help distinguish the utterings of timid speakers in noisy environments.

At home, these tentacles will be connected to your bathroom scale and sink, your refrigerator, and many other electronic kitchen appliances, especially your kitchen's special autocook facilities for automatic preparation of light meals. In your living room, the tentacles will connect to your stereo receiver, TV, VCR, massive info jukebox, and the rest of your entertainment equipment. Your home E21 will also be connected to your heating, air-conditioning, and sprinkler systems, as well as your phones, lights, and many other devices. Wall-mounted, touch-sensitive displays with microphones, speakers, and cameras will be connected to your E21 to help you interact with your system, when and where you need to. I have already calculated that I will need 14 of these ports in my home: three in the kitchen (for planning, cooking, and near the table where we eat), one in my study, one in the workshop, two in the living room, two in the bathroom . . . you get the idea.

In your car, the E21 will ride in the trunk, with speakers that give you synthesized speech messages and microphones that listen to what you say. As driver, you may see a heads-up display in your windshield, while passengers may get regular pop-up screens. Your car's E21 will be able to communicate with all cellular and wireless networks wherever you are going. Its tentacles will link cameras that can look outside, environmental controls, and more mundane controls for windows, seats, and mirrors. This way, by just speaking out loud, you'll be able to turn up the heat, switch to a different radio station, inquire about the traffic patterns ahead, find out if it will rain tomorrow, send a message, ask and get directions, and take a digital picture of a passing car with a raging driver—all without taking your eyes off the road or your hands off the wheel.

The N21 Network

Network 21 is a set of network protocols—agreements and conventions about communicating information among systems. The software that implements these protocols sits inside every H21 and E21. Its purpose is to help the H21s and E21s cope with mobility, interrogate physical devices, form secure collaborative regions, communicate over different networks as needs dictate, and adapt to changes in the communications environment. The N21 is not built from scratch. It is an additional set of capabilities on top of the protocols that handle the Internet and the Web.

A new network protocol is needed because existing networks, including the Internet, were not designed for mobile users, but rather to interconnect big clunky computers that were expected to stand still. Internet computers are identified and located by a so-called IP address, such as 18.49.1.200, or its more readable name—hq.lcs.mit.edu. One of the key goals of Oxygen is to help people and computers easily discover and access devices, services, and information, even when the people or the devices are moving around. In N21 this is done with a scheme called Intentional Naming System, the brainchild of Hari Balakrishnan at LCS.

To address a resource, you specify a property that you want the resource to meet—for example, "nearest uncongested printer." The N21 approach uses a clever electronic location support system called "Cricket" to help your machine locate the resources you may need, without giving away your own location. The resources periodically broadcast a simultaneous radio frequency and ultrasound signal that your H21 can sense. Because radio waves travel at the speed of light, your computer detects this signal immediately as it is broadcast. The ultrasound pulse, on the other hand, is detected later, because it travels more slowly. The time delay between the two signals tells your computer exactly how far each resource is from your machine, even as you move around. Using the Intentional Naming System, your machine can also find out how busy the various resources are, so it can complete the task of finding the nearest uncongested printer. In this way, the address of a physical device you may seek, or of a person's portable, is not fixed, but is resolved at the time you ask for it, and can change even during the course of a conversation.

Being able to access physical devices by "intent" is very important in human-centric computing because we will be surrounded by thousands of devices, and keeping track of them only by their physical address would not work. The astute reader will have noticed that this is yet another manifestation of our favorite ascent toward meaning—

you don't just give the location of a resource to which you wish to become interconnected. Rather, you say something about what you want to do with that resource: "Let me talk to Michael," "Show me what's happening in Office 516," "Get me the temperature controller for this room." This capability is at once important and dangerous. It's important because it is natural and conveys the meaning of what we want to do. It's dangerous because our "intent" may be too complicated and bog down the N21, which may be unable to resolve it—for example, if I say, "Get me a camera near the person responsible for issuing my passport." Intentional naming will be most useful if it is used "thinly" to resolve simple intents, like locating a person by name, or a device by the direct function it performs.

Another important property of Network 21 is its ability to adapt to a variety of changing communications conditions, which we humans will impose as we run around. When you are near your office and use your H21, the wireless signals will be strong because of the nearby powerful E21s and because your portable won't be stingy with its own power, since it knows it's near a charging outlet. But when you are far away, signal strength may vary unpredictably from one moment to the next, since radio frequency signals are affected by terrain, weather, and electromagnetic interference caused by other transmitters. To manage changing conditions, the N21 uses a technique called "Radio-active Networking," through which the communication protocols and transmission methods adapt on the fly, using a range of sophisticated adaptation techniques. They might change data rates, change channels to avoid noise, and replicate signals that get corrupted, allowing applications to adapt what they send to current network conditions.

The N21 is adaptable in other ways, too. For example, people who use cell phones are familiar with the "horizontal handoff" that happens when they are moving (typically, driving) from one calling region to the next. Ideally, the stationary antenna of the "cell" they are exiting drops them just as the antenna of the cell into which they are entering picks them up . . . without losing any part of the communication. In the N21, such handoffs will take place as you move horizontally among different regions of the same network. But a new kind of "vertical handoff" will also take place when you leave one communication regime, like a high-speed network near your building, and adopt another communication regime, such as a cellular network. The horizontal and vertical handoffs bring the sophistication of actively changing communications conventions to a higher level—a new territory to be mined for its capabilities and problems.

Another vital adaptation capability of the N21 is rapid self-organization of a handful of H21s and E21s into a secure collaborative region. You and your collaborators won't have to do anything other than set a level of privacy, and the degree to which hyperfiles can be shared or edited, and your N21 software will invoke the appropriate security measures to make this possible. Several such secure collab schemes will be present already in your system for your use, such as "personal," "group confidential," and "company wide." When you are finished, the N21 will ensure that the secure collab region is demolished, just as rapidly as it was formed.

The Oxygen network will also have to adapt to the different speeds of the signals generated by physical devices and appliances. A room thermostat is a very slow device that may need to be read only once a minute, whereas a stream of data coming from video cameras tracking the comments and gestures of participants in a meeting requires a network speed that is a billion times faster. The N21 must be able to adapt to these variations, and it must also support the nomadic software flows that give its devices different capabilities, upgrade user software, and cope with software errors.

All these demanding requirements must be handled if Oxygen is to meet people's needs.

Speech

The Handy 21, the Enviro 21, and Network 21 are the main delivery vehicles of the Oxygen system—and are fairly easy to visualize. The Oxygen system software that supports the hardware is a bit more abstract, and is best described by how it handles each of the five human-centric forces.

The H21 has no keypad, so if it can't handle speech, the Oxygen users would suffocate. Initially, Oxygen will use the "light" speech systems—those that work in a very narrow context and carry out limited tasks, such as controlling a piece of equipment or fetching a document. Each application has to provide descriptions of the expected speech usage. These descriptions are fairly short. For example, if an application wants to let you control your radio, it would provide Oxygen with a description that, slightly paraphrased, would look like this:

Speech Module Name: Radio Controller

When I say "on," or "turn on," or "blast"; Produce as output: "Radio-On"

When I say "off," or "turn off," or "kill"; Produce as output: "Radio-Off"

When I say "louder"; Produce as output: "Increase Radio Volume by 10 percent"

When I say "softer"; Produce as output: "Decrease Radio Volume by 10 percent"

When I say "tune in WBUR"; Produce as output: "Set frequency to 90.9"

Oxygen has a speech understanding compiler—a program that uses a description like this to generate a light-speech software module, which in this case is called "radio controller." The module becomes part of the application and is fired up whenever the application runs. If you then say, "Would you please turn on the radio?" the radio controller module will take in your spoken phrase and will produce the signal "Radio-On." This output signal will then be fed through an Oxygen automation module to the electronic equipment to turn the radio on. What happened here is that the Oxygen compiler, based on the description, generated a mini speech understanding system that can comprehend the specified key phrases and respond with the outputs that correspond to these phrases. The compiler imparts a lot of its knowledge about speech to this module. For example, it endows it with the right software to disregard filler words, inverted phrases, and other common speech idiosyncrasies that are extraneous to what you want done. The module concentrates on the few things it is expected to understand and listens for them with a keen ear.

Now imagine that you say, "Power up the radio." The radio controller module will complain, saying, "I don't know what 'power up' means. Please explain." At that point you could say, " 'Power up' means 'turn on,' " and the system would add this new key phrase to the radio controller speech module. With such additions, a speech module enriches its capabilities, adapting to the phraseology of its user for future action.

Let's assume that the same "home comfort" application also lets you control the climate of your living room. The application programmers would have provided a description of the expected speech usage for a "climate controller" that sets the room temperature and humidity.

As you give a spoken command, Oxygen needs to understand whether you want to control the radio or the room temperature. It does this with an additional piece of software called the "speech switch," which is always listening and acts as a policeman that routes commands to the proper speech module. Normally, you would first address the speech switch by saying "switch to climate controller" and the system would confirm by repeating, in a pleasant voice, "climate controller." After that, your spoken commands would be directed by the switch to the climate control speech module, until you speak the name of another module.

That's the "safe bottoms" way of switching contexts explicitly. But we don't want to have to be so self-conscious about what we are saying—and we need an unsafe challenge that would eliminate this awkward crutch. So the Oxygen plan also calls for exploring implicit switching. If you say, "Please turn on the radio," the climate controller will issue the "don't understand" output while the radio controller issues a valid output message. Each module also issues confidence numbers that measure how well it thinks it has understood the spoken phrase. At that point the speech switch, observing these results, would decide that the message was headed for the radio controller rather than the climate controller, and would automatically switch control to that module. But if you mumbled, "I want it hot," both modules might respond, with lower confidence numbers, and with the music controller thinking it heard "I want it off." At that point, additional context information could be brought into action through the semantic Web. For example, if you had just asked about the room temperature, the system would determine that you are probably more interested in adjusting the temperature, rather than the radio. Of course, you may always revert to explicit switching, much like you would toward a person who becomes confused by something you mumbled. This is all new territory. We won't know how far we can go with implicit switching until we try it.

After using Oxygen for a while, it's possible you would end up with too many speech modules in an application. Unable to remember what is available, you might ask, "What speech modules do I have?" and the system would speak or show them to you.

Normally, the descriptions of Oxygen speech modules would be written by applications programmers. Oxygen would also have several speech modules that would understand commands shared by all applications, like "Please show the results on the screen."

But the process of introducing a new speech module should be easy enough that you could also do it by yourself. We hope that with increased usage and additional technology we will develop, each light-speech module will grow in sophistication and gradually become a full-fledged narrow-context system, like the Jupiter weather information system.

Vision will be used in Oxygen for identifying people by camera, and to reinforce speech understanding; for example, by showing images of objects in a meeting room. The software approach for these uses is similar to that for speech—through vision modules. Additional multi-modal procedures will be provided by applications programmers to detect when a speech module and a vision module reinforce each other's message, thereby making it easier to understand a user's intent.

Automation

Oxygen handles automation similarly to the way it handles speech—by generating automation modules against short descriptions of the tasks these modules are expected to carry out. Hundreds of these "automation scripts" will be included as part of applications programs. Software developers will supply these scripts individually, or in bundles such as "house" scripts or "office" scripts, or will make them available through large, customized applications packages to hospitals and financial institutions. And, as in speech, you will always be able to create custom scripts of your own.

Automation scripts are written in an English-like programming language and specify how certain information-processing tasks should be related to each other. For example, the task of alerting you if an e-mail message or phone call from Joseph Bitdiddle comes to your office would be described in a script like the following:

Automation Module Name: "Bitdiddle Alert"

If incoming caller id is phone # xxx-xxxx or if incoming e-mail "sender" contains "Bitdiddle,"

then . . .

The first line tells Oxygen this is an automation script and gives it a name. The next line grabs the caller ID of incoming calls and the sender's name of incoming e-mails and checks to see if either of these two pieces of information matches Joe's phone number or e-mail name. If there is a match, then the module goes on to call you or alert you in whatever way you specify. The automation compiler of Oxygen uses this script to generate an automation module—a software program that runs whenever the application that introduced the script gets fired up. In this example, the application is "on" all the time, checking all incoming phone calls and e-mail messages.

Creating automation scripts that monitor incoming phone calls and e-mails will be such a common activity that a speech module that lets you create such scripts with spoken commands is sure to be part of the basic Oxygen system. For the above example, you would say to the Alert speech module something like this: "Create script to alert me if Bitdiddle calls or e-mails me." The speech system would then translate this statement into a script like the one above. Once you confirm the procedure, Oxygen would pass the script to the automation system, which in turn would generate the automation module dedicated to doing this job.

After some use, your Oxygen system would accumulate several automation modules that would be activated with each application you run. At any time, you could ask to see all the associated scripts so that you can edit them, eliminate them, or add new ones.

Automation scripts will often be created to command physical devices and appliances. This means that for every connected device, there must be a corresponding software module to translate commands into actions. Conversely, Oxygen will have to convert the physical states of devices into information the software can use. A simple example is a garage door that can be electrically opened and closed while you are at work, say, to let in a deliveryman. Sensors would have to be mounted that can ascertain the position of the door and communicate it to the home E21. After you saw a video of the deliveryman's face and spoke to him, you would issue the command to the automation device module that operates the garage-door opener.

These device modules are the software representatives of the physical devices. To do anything to a device, you must tell it to its device module; and to find out anything about the device, you must ask its module.

The prospect of creating an automation script for every movable, electrical, or digital device in your home—not to mention installing sensors all over the place that communicate over radio waves to your E21s—might seem daunting or not worth the effort. If we have to spend days creating scripts for every last little action we want Oxygen to take, or hundreds of dollars for sensors, we would still be serving the machine! But that's not what will happen.

First, you will not connect every conceivable device just to be faddish. You will focus only on the ones you really need to control. Second, vendors of physical devices and suppliers of home automation systems—a new industry that is already emerging—will provide built-in scripts that will handle typical functions. At work, companies will create all sorts of useful scripts for their employees, en masse. Hospitals will do it for their doctors and nurses. As for rewiring the house to carry N21 to all your sensors, you won't have to. Radio frequency systems that operate within your home will be able to handle all the necessary communications. One example is the emerging Bluetooth standard, which provides wireless communication among devices within about 30 meters.

When the newness of interconnected physical devices subsides, the effect will be very much like that of having motors in your home, car, and office. You will hardly bother to know where the motors are located, or think twice about how to control them. You will just use them to serve your needs.

Oxygen, however, would be aware of all the details, for they are an important part of the chain that brings the physical and information worlds together. Physical devices can have all sorts of different electrical characteristics; some are activated by high voltages, or by pulses, or by being given a number. When they report back what they perceive (like the room temperature), or what their state is (like an open garage door), they may do so in any one of these different forms, at various speeds, with varying rates of data. A configurable electronic interface card has been devised by Srini Devadas, who heads Oxygen's automation effort, to take care of these details for a wide range of appliances ranging from simple motors to sophisticated computer peripherals, converting each appliance into a software device module that can be handled by Oxygen's automation software. We can be sure that many standards will emerge toward the same goal.

Individualized Information Access

Software to handle individual information access is being designed as this book is being written, so I'll speculate a bit on how it may turn out. The foundation is a new Web-like "file system" that lets you organize and thread information with the same meaning—the concepts that are important to you—under what I have been calling semantic, or "red," links. Oxygen lets applications developers—and you, the user—create the initial concepts of interest to you, as your individual system's top-level links. You see these concepts as clickable Web links that access information inside your computer. Oxygen also offers a convenient way to organize and navigate through these links. Various schemes are under consideration, but for now, assume that Oxygen will use the virtual geographic metaphor where information is organized in two-dimensional maps, according to its meaning.

Oxygen also lets the applications developers, or you, provide links to important databases that you will be using often, regardless of whether they are inside your organization or outside, on the Web. These pointers will be organized and threaded under the same red links that specialize the system to your purposes.

So if you are a doctor, your Oxygen system will come to you with your "clinic application" preloaded. Turn it on and you see a map of different cities. To the west are disorders and illnesses, with names like "Lung disorders," "Hormonal disorders," "Digestive disorders." Click on one and you see, organized in different streets and neighborhoods, subordinate disorders, with symptoms, diagnosis, and treatment information. To the east you see cities with names like "Children's health issues," "Drugs," "Patients." You click on "Patients" and see your patients, each in a building, with a little photo on the roof that identifies each person. You click on Mr. Jones and see in the building's different rooms the different kinds of information that is accumulated for that patient: test results, visits, comments of doctors, prescriptions. Every patient is in a similar building. After a while you will be navigating easily through these familiar rooms, but it is the cross-referencing that catches your attention. While you are in Mr. Jones's building, you can ask for Medline info on his illness, or for any other patients with the same illness, or for drugs indicated for that illness, and so on.

With many important concepts preestablished, the automatic parts of the individualized information retrieval can now come into play. Oxygen uses several sophisticated programs, based on the Haystack technology, that act as observers and extractors. Here is a very partial description: Any text you touch is examined for title, author, date, and other such metadata, which is extracted and tags the document. The document is then stripped of all format niceties and reduced to stark text. All common words, like "and" and "have" are thrown away, and an inverted index is formed of the document's keywords. Metadata is compared between the document you are looking at and other document headers in your various red-link categories. Similar comparisons are carried out by counting the number of shared words in the inverted indexes. When the results of these comparisons show a tight correlation, the document you are looking at is automatically attached to the red links of related documents. When Oxygen is not so sure of the relationship, it will put its quandary in a bin where you may later advise the system as to which semantic link this document best belongs.

The other major piece of software in the individual access part of Oxygen is the query system. You may, of course, use speech to ask questions, which is the task of the speech understanding system. But you will also want to input complex queries, using a scripted language similar to that used in automation. The answers will be links to documents, the documents themselves, and sections of documents. The query system works very much like a traditional information-retrieval system, except that it is now acting on red links that are close to your interests, including your own links, those of your associates, red links on the Web, and, last, plain old-fashioned blue Web links. The search process is semiautomatic, in the sense that if Oxygen can't quite get you there, it lets you chime in with your navigational smarts.

Collaboration

The heart of Oxygen's collaboration software system is the collaboration editor. Its primary capability is to use meaning to thread together the goings-on in a meeting. Its built-in concepts include "meeting," "location," "participants," "discussion topics," "open or closed issues," "conclusions," "information used," "simulations," "summaries," "conversation fragments," and a great deal more. As application developers and users further tailor Oxygen to individual specialties, they will add customized red-link categories to these generic concepts.

Shared with N21, an important function of the collaboration software will be to help form and manage secure collab regions. In a one-on-one synchronous encounter, or a group meeting spread over space and time, or a theater-like event, the participants will agree to certain rules about the access and control rights they will have over the information they examine, generate, and modify. Whereas N21 deals with the mechanistic aspects of setting up these regions, the collaboration software focuses on the human and machine procedures that will make formation of these regions easy and acceptable to people.

With such collaboration software in place, Oxygen applications programmers would make different collaboration templates available, say, for insurance and finance firms for processing claims, applying for loans, and the like. These templates will be used by local and distant workers to proffer their information work, following the style and format that is suitable and convenient for each individual category of work.

The safe way of using the collab editor is to have it keep track of routine things, like document versions, and have a human secretary, perhaps a meeting participant, oversee the pulling together of all the various pieces in collaborative session hyperfiles. The combination of these simple capabilities can go a long way toward handling the human-centric needs of people who work across space and time. It is through the judicious use of specialized templates and hyperfiles that a great deal of information work will take place on Oxygen.

More challenging activities are being pursued by Mark Ackerman, Trevor Darrell, and Howard Shrobe of MIT. They involve expanding the collaboration editor's capabilities to model what is going on in a meeting and to mediate the interactions among the participants. In modeling, the editor will use speech understanding, and track the interaction of users with specific kinds of information and with simulations, to identify participants who propose positions, make arguments for a particular position, or who draw conclusions. The editor would then categorize these contributions and thread them for future access. The researchers believe they can make the editor deduce the principal flow of a meeting without really "understanding" what is going on, but rather by acting on significant clues. Of course, such deductions will be imperfect and will result in errors. But they might form an adequate first cut at organizing what happened in a meeting that would then be further refined by a human secretary. And that could help organizations hold more effective meetings.

The mediation function of the collab editor would require that it be given enough information about the work flow and the goals of the participants to be able to check if the right steps are being taken. We saw this in the example of collaborative car design, when the collab editor caught a constraint violation and alerted the designers. It is possible to lay out certain checks that the collaboration editor should perform to ensure that all the right actions are being taken along a prescribed work path. This elevates the role of the editor from a mere recorder of activities to one that participates and coordinates, however modestly.

Customization

Customization is not an independent technology in Oxygen. It is a property embedded by applications developers and users into Oxygen's speech, automation, information access, and collaboration technologies, and in the applications written for it. Customization continues throughout the lifetime of an Oxygen system as developers issue upgrades, and as individual users add their own scripts and preferences.

Taken together, these customization characteristics form a user's "info personality," which will change with time. The Oxygen system ensures that anonymous H21s and E21s can be personalized with these characteristics, thus customizing the hardware to individuals and their needs.

The Oxygen Software System

The Oxygen "software system" ties everything together. It is made up of a User Operating System, a Machine Operating System, and a Bridge Operating System. Copies of this software flow, as needed, in every H21 and E21.

The User OS is the applications interface for speech understanding, automation, individualized information access, and collaboration. And it contains the meaning-oriented hyperfile system of Oxygen, which each application can call up and customize. The User OS is where people interact directly with the machines and each other, through Oxygen's top-level metaphor (not yet chosen), in the same way you interact today through the desktop metaphor of Windows or Mac OS. In addition, each application plugs into Oxygen through the User OS, by supplying scripts and templates for each of the human-centric technologies and for the ways they should be combined with one another.

Oxygen's Machine OS is very much like today's operating systems, without the desktop icon-and-mouse interface. It is the collection of low-level machine calls and directives, like information transfers, copying, naming and renaming, connecting to communications sockets, generating text and graphics, and the many other things machines need to do at their level. To carry out these actions, Oxygen uses parts of a conventional operating system called Linux, a variant of Unix that has the additional good property of being "open," meaning that we can modify it as we wish.

The User OS of Oxygen caters to what people want to do. The Machine OS caters to the low-level actions machines need to do. The third piece of the Oxygen software system, the Bridge OS, ties these two parts together. It translates the actions called for by the User OS into Machine OS commands. It implements the secure collaborative regions, and all the N21 functions of address resolution, vertical and horizontal handoff, power-communications-computation trade-offs, and so forth. The Bridge OS is also responsible for providing the management strategies that move the Oxygen application and system software seamlessly among E21s and H21s. This requires an organization of all the software into small software objects, sort of like little balls, that can easily roll from one machine to another. The Bridge OS also provides the "eternal" property of all Oxygen software, which ensures that you never have to reboot your machines. This minor miracle is accomplished with so-called checkpoints: If your system runs into trouble, it reverts back to its most recent trouble-free state, from where you can move forward once again—techniques well known in the world of large databases that cannot afford to crash.

Turning on a Dime

The big claim behind human-centric computing is that by focusing on a few technologies that are close to what people want, we can come close to offering the gas pedal, brake, and steering wheel of the Information Age.

The proof will be in the prototyping of a human-centric system, whether it's Oxygen or some other approach. The first crude Oxygen prototype won't be ready until 2002, and then it will only be available for internal testing by MIT and its partners. That's when we'll pull together the various pieces, mostly through their safe incarnations, to see what Oxygen can really do for us. In technical parlance, this is where we will explore the integrative potential of Oxygen, which, if we are right, should go well beyond the power of its individual technologies.

The tests and demos that we will carry out on this prototype will affect significantly the direction of future prototypes. A key application of the Oxygen prototype system is in health care, particularly in implementing the Guardian Angel application. We also intend to use Oxygen in our daily lives as we build it, and to build it as we use it, based on our experiences. Somewhere in this process of successive improvements, backtracking, and abrupt turns we will move to different, more streamlined hardware and software, based on what we have learned, which we will deploy in the hundreds among our researchers. This process will eventually culminate in our final prototype (a most dangerous statement in the world of software) some time around 2004.

Assuming success, the availability of Oxygen to the world will depend on how it will actually end up being disseminated. The basic Oxygen software will be freely available to anyone, in the public domain, without any restrictions whatsoever. It might be commercialized by a start-up company, which would have to raise considerable capital. Or it could be suddenly embraced by a giant, like IBM did with Unix and Microsoft did with the Web, causing a huge company to turn on a dime and "go human-centric." Then again, there is the Oxygen Alliance. This group is interested in the pursuit of pervasive human-centered computing. They have many systems and resources of their own that could make a prototype and an early dissemination of Oxygen and its techniques a reality.

Granted that Oxygen or some other system reaches the human-centric level that helps people do more by doing less. What does that really mean? An increased human productivity together with greater ease of use and a lot of fun? Certainly.

But could it mean more?

Nine

FINISHING THE
UNFINISHED
REVOLUTION

Imagine it's the year 2020 and the radical change we are after has happened. Systems like Oxygen have finally risen above the machine level and have been serving human needs. How far have they gone toward helping us do more by doing less? Did they help us get rid of the many difficulties that surrounded computers back in the year 2000? Did they increase our productivity and make our systems easier to use?

Back at the turn of the century, we had to read huge manuals to operate a word processor. Now, thanks to the natural interaction provided by human-centered systems, this "excessive learning fault" is largely gone. We talk to our systems and they understand enough to talk back and be useful. We still have to learn how to operate these machines, but the effort required on our part is much smaller.

In 2000, we typed and squinted a lot, doing all the electronic shoveling with our brains, eyeballs, and fingertips. Human-centric automation has freed us from this "manual labor fault," carrying out all sorts of tasks automatically. The "human servitude fault" is also largely behind us, since in the face of truly useful automation, service providers can no longer get away with those terrible automated phone operators that enslaved us through a maze of push-button choices.

The "overload fault," caused mostly by a dangerously expanding e-mail habit, has also been brought under control. People have adopted human-centric attitudes—they no longer frantically send so much unsolicited e-mail, nor do they feel obligated to respond to every message they get. Most legislatures have passed laws obliging tele-marketers to tag their messages with metadata that identifies the sender and the category of product or service being proffered, and filters used by essentially everyone let through only the ads that their masters wish to see.

Before human-centered systems we could barely find what we wanted through all the info-junk. Today, even though the info-junk has soared, we can find what we want with less work on our part, thanks to individualized information access and the ascent to meaning through the Semantic Web; the old "information access fault" has been largely circumvented. The "feature overload fault" is also out of the picture, because customization of our systems and applications to our individual needs have reduced the tendency of software developers to provide every conceivable feature in an attempt to please everybody. The old "crash fault" has been vanquished, too, because our human-centric software tracks the daily evolution of every program we touch, bringing us its most recent incarnation, and because when we run into trouble, the system takes us back to the most recent trouble-free state. Our machines do the backing up, not we. And no longer do we have to contend with the loss of time and peace of mind to port our software from one machine to another when we change machines. Nomadic software ensures that our info personality flows into whatever new piece of hardware we acquire or borrow, wherever and whenever this is necessary.

The "unintegrated systems fault" that made it impossible for me to use my calendar card during my plane ride to Taiwan is now a rare occurrence. The human-centric focus of technology has made the developers of operating systems and applications much more conscious of the need to serve people, and competition to supply consumers with this highly desired, higher level of operation has obliged them to do so.

Not all the computer faults have vanished. The "fake intelligence fault" continues to bother us, as software developers try to make systems more helpful by making them more "intelligent." And the collection of hundreds of automated procedures that we all have, while helpful in their individual tasks, conflict with one another at times. The "ratchet fault" where layers of old software pile up on top of one another is also present, because writing software continues to be more of a difficult arts-and-crafts proposition than a precise science, and we have not yet come up with any dramatic improvements to the software design process. We derive some comfort, however, from the fact that most of this ugliness resides well inside our systems, invisible to us.

In 2000, we all plied our trades and pursued our private info escapades with identical "personal" computers. Today, the machines adapt to our unique needs through customization. Back then, we could not easily reach people on the go, nor control our physical surroundings. Now we use the human-centered systems' ample reach to interact with people in every place and time, and to control the devices and appliances we care about.

Human-centered systems also have made it possible for us to carry out new tasks. They help us work easily with one another across space and time, tracking our activities, helping us form secure collaborative regions, letting us annotate our conclusions, and generally helping us work much better than we could using only e-mail. Information work is now routine and occupies one-fourth of the world's economy, as people buy and sell human information skills across the world. Nearly 10 percent of that activity comes from India, which has doubled its GDP since 2000, mostly by selling clerical office work and software services. China accounts for 6 percent of total information work and

Africa for 3 percent. The remaining 80 percent is within the industrial world.

Of the 1.5 billion people now using the Information Marketplace, some 300 million come from these three vast blocks of humanity—a feat made possible with a lot of good help from the people of the West, and partially from the progress in cellular Web access and in speech technology. Our goal of ensuring that many people become interconnected has been partially met, though we are by no means there. Compared with 2000, when fewer than 5 percent of the world's people were interconnected, the figure now approaches a respectable 20 percent, a quarter of which represents the developing world.

A principal objective of human-centric computing was to develop the gas pedal, steering wheel, and brake of the Information Age. We have done so in the form of the five human-centric technologies, which became the applications interface of our new systems, were adopted by a new breed of applications, and have sent our productivity soaring. And by infusing these technologies into the Internet and Web, we have transformed these old media from their "voyeurism and exhibitionism" state into a full-fledged Information Marketplace.

Information technology has come well into our lives, and, as expected, we notice it less. Human-centered systems have liberated us from thinking about technology to thinking about what we really want to do. We can rejoice in the knowledge that our beastly computer menagerie of old has been almost fully domesticated!

But we are not quite where the Industrial Revolution was in 2000, because our information systems have not vanished as completely as the motors had back then. More work will be needed as human-centric information technologies continue to penetrate new areas of our personal and professional lives. When our information systems finally vanish in another decade or two, that will be the signal that the Information Revolution is done.

Let's suppose that these estimates for the year 2020 or so are correct, and the Information Revolution has been finished in the same sense that the Industrial Revolution is now over. Will we then be better off? Or will we have become efficiency freaks, bent on being productive every moment of our lives, in the process losing our peace of mind, our humanity, and our heart and soul? What will we do with all the time we save—work more? Will computers that operate at a more human level help us be more human? Or will our increased preoccupation with information drown us? Will increased collaboration across the planet lead to a uniform global culture? Will automation and superior information access make us lazy and excessively dependent on our machines? Or will the new capabilities encourage us to follow the high road? Will simpler systems reach beyond the fraction of the globe they now serve, to the billions of still unconnected and unengaged people? If so, will the systems help poor people become wealthier? Ultimately, how far might we go with human-centered computers toward enhancing our humanity?

Let's get some answers.

Info Royalty

We begin our search for the big picture with a small step: What might human-centric computing do to our rational, utilitarian selves? The answer is straightforward, especially in comparison with the Industrial Revolution. If you like what cars, airplanes, electricity, and chemicals have done for you, then you will like what the new information tools will give you.

You will be able to do more work, especially of the office variety, in less time. You won't be as frustrated, because your systems will be easier to use and more responsive to your needs. Your health will be improved through less expensive but faster, more accurate, and higher-quality medical systems. And you will have faster access to more of the world's products and services, tailored to your special desires. All the services you normally use, from getting an appliance fixed to finding the right lawyer or a comfortable future home, will be faster and better. You will have more options on receiving instruction, and even becoming educated. New entertainment will surround you, rich in content and interactions with other people. And you will have fun in new ways, as you play with it. You will also interact more easily and reliably with your family members, wherever they and you may be. Your thoughts and ideas will touch more people, and you will have the option to visit more of the thoughts and ideas of your fellow human beings. Organizations will function more efficiently, too, including governments, which will be able to better reach and interact with their constituencies.

These utilitarian benefits are qualitatively similar to the benefits we and our ancestors received from the plow and the motor. These earlier tools helped increase human productivity dramatically. Nowhere else is their combined effect more visible than in the generation of food, which went from absorbing all people in ancient times, to occupying a mere 2 percent of the industrial world population today—a whopping 5,000 percent productivity increase. These industrial innovations also helped us live better and have fun in new ways, with bright lights, automobiles, aircraft travel, consumer electronics, useful medicines, and so much more. As with human-centric technologies, these industrial developments made it easier for people to carry out their professional and personal lives. Just compare all the personal and professional things we can do today using the automobile with what people could do in earlier times using their feet and an occasional horse.

Of course, you might argue that cars did bad things for the family, the environment, and our soul, or that factory automation displaced jobs and led to the atrophy of our muscles. The same scenarios will be repeated in the Information Age.

How about leadership, responsibility, honesty, and those other human qualities we treasure? The answer is that you'll be able to use the new tools to either further or diminish these qualities. Any change will be up to you. As for the new ills that human-centric computing may bring—theft at a distance of our money, sexual advances toward our children, misinformation about us, cross-border crimes—the same answer applies: The new tools, like all technology, can and will be used for good and for evil. The angels and the devils are not in the machines, but in you and me. Since the ratio of angels to devils stems from human nature, this proportion is not likely to change. The balance between good and evil in the world won't be affected by the onset of human-centric systems.

Almost all the arguments you can fashion today about what the world of information will do to us were raised during the Industrial Revolution. So ask yourself if, considering everything you care about, you are better off with that socioeconomic movement behind you. Or would you be happier if it never happened? With almost no exception, the people of the industrial world have elected to live in it rather than in a cave, foraging to feed their families. This suggests that despite protestations here and there, people overwhelmingly prefer the industrial to the preindustrial way of life.

I can already hear the dissonant chorus: "People can't help it." "They think they are better off but they aren't." "This is a utilitarian society that has lost its compass heading. No wonder they like it. They have lost their sense of direction." I'll address these deeper questions about technology's ultimate impact upon humanity in a moment. Meanwhile, it is safe to conclude that from a utilitarian perspective, we will be better off with our new information tools, for the same reasons that we continue to be satisfied with the greater utility made possible by the industrial advances of the previous two centuries.

Does all this mean that human-centered computers will simply continue the same sorts of benefits? Not quite. The gains will be sufficiently different to induce a qualitatively new social change—something akin to a new social order. In a strange way, we'll be able to do many of the things that were the province of wealthy people, past and present. Kings and rich folk have always had servants that catered to their every wish. With human-centered computers, we, too, will end up surrounded by many automated servants—scripts and specialized procedures ready to cater to our needs. Rich people have always had better access than the rest of us to the information they need, because they have the right connections and can afford the expense of finding and obtaining what they need. So will we with the human-centric force of individualized information access. Rich people have always had products and services customized to their desires. So will we through customization. Rich people don't need to work, because their wealth breeds more wealth. This won't happen to us completely, or overnight, but the expected threefold increase in human productivity, made possible by human-centric systems, could free up two-thirds of the time we now spend working . . . if we elect to realize the savings in this way. The collective benefits of human-centered machines will give us enough of the capabilities now reserved for the rich to make us feel like royalty.

Just as the Industrial Revolution produced a new middle class, the Information Revolution, through it human-centric technologies, will create a new "info royalty" class. Who knows? A few decades from now, human-centered machines may return human beings to the princely benefits of earlier feudal times, when the rich had servants, and the master reigned supreme . . . except that almost everyone will have a chance of being the master!

Will we then be better off? That will be up to us. The history of kings and princes shows that they have gone in every conceivable direction during their spare time. If we follow in the Information Era what we did in the Industrial Era, we'll work harder with the time saved by our new royal status, so that we may acquire even greater wealth. On the other hand, we may elect to devote the time we save to other endeavors that please or uplift us or benefit those who are less fortunate. We'll have the luxury of choosing our course.

Such a societal shift would be more profound than an incremental utilitarian improvement in human productivity and ease of use. Doing more with less effort would then have the added meaning that we would be able to act more like kings than serfs.

Global Reach

Who would have believed 15 years ago that poor programmers in Bangalore, India, would sell their software services to the West, putting together companies like Infosys (which in July 2000 was valued at nearly $25 billion), which collectively employ 60,000 programmers, whose standard of living is now pulling their region's economy upward at 25 percent per year?

That ray of sunshine is particularly important for the hope it brings to the developing world. For, if the new royalty class is limited to the people now interconnected via the Web, humanity won't be doing more by doing less. The new "royalty" would stand for a tiny fraction of the world population. And that would be just as bad as the real royalty of old, reverting us to an era of a privileged few, likely to be followed by bloody revolutions, as was feudalism. This is why I insist that a primary imperative of finishing the Information Revolution is that the new technologies of information reach as many people as possible.

Fortunately, there are many ways to improve the global reach of information technology. Communications could be provided by low-earth-orbiting satellites operated by such companies as McCaw Communications and Globalstar that whip around the earth. When these birds are over the industrial nations they are very busy, but when they are over the developing world they are doing nothing. Let's pay the low marginal cost to leave them on. In addition, hardware and software makers, training outfits, and communication service providers could offer their wares to the poor at deep discounts. We citizens could help cover the cost by instructing our governments to offer attractive tax breaks to these suppliers. Individuals could also donate money or time. Organizations like the World Bank, which spends over $30 billion annually in structural loans to the developing world, could put a good part of these funds into worthy information technology projects.

Armed with the excitement of these prospects, a few of us techies got together with a colleague from Nepal, fully expecting to boost his nation's economy by 20 percent through clever use of the Internet. Unfortunately, we quickly found that even if we got him the communications, hardware, software, and training for free, we would still fall short of our goal. That's because only 27 percent of the Nepalese are literate, and of those, only a small fraction can handle English. When we asked what services that smaller group could offer we hit a brick wall. Many are not skilled, and those who are, are busily running their nation's businesses. Maybe we were too ambitious when we envisioned a future workforce in Nepal selling office services to New York and London via the Web. The potential of the Information Age seemed overshadowed at every turn by the ancient forces that separate the rich from the poor.

Like others who have tried to do something in this area, we, too, came to the realization that the lack of communications, computers, and training is not the primary problem. The bigger obstacles are the same that have kept the poor from rising above poverty throughout history. Lack of education is at the helm. It is followed by lack of transportation, power, and telecommunications; absence of capital; misuse of whatever resources may be available; government inertia; and cultural taboos. Moreover, basic concerns over food, shelter, and health dominate poor people's plans and actions, as they should, ahead of the less tangible promises of information technology.

These observations and concerns were amplified by an MIT Laboratory for Computer Science survey about the uses of information technology in the developing world in 1999. The results showed that the biggest recent successes in developing countries, disguised under all sorts of information technology experiments, actually involved the introduction and use of POTS—plain old telephone service. And in cases where new information technologies beyond telephony seemed to be statistically active, we found that they were used mainly by the few relatively rich people among the poor—a faithful microcosm of what is happening globally, and hardly a model for addressing the larger problem. We have not yet latched on to an approach that can productively engage the poor in the global Information Marketplace.

If the world has to hold out until developing nations, and the poor in the industrial world's inner cities, fix in serial fashion the social, political, and economic problems that plague them, we will be in for a very long wait. What we must do instead is help through donations, government aid, personal and corporate contributions, tax credits, loans, and all the mechanisms we can muster to improve education and infrastructure. Most important, we must explore creative "shortcuts" that have a chance of working. One possibility is to strengthen entrepreneurial initiatives among the poor through incubator programs that provide capital and other resources. Successes from within a community, as in the case of Bangalore, will stimulate duplication far more effectively than solutions from outside. Another shortcut may be the launching of short-term training and education programs aimed at preparing people directly for selling information work. Yet another shortcut involves the use of speech understanding technology to bypass illiteracy for people who, despite their inability to read and write, can contribute and benefit from the Information Marketplace.

A new world of human-centric computing must work for all humans. If the bulk of our planet's people are not interconnected, then humankind will not be able to do more by doing less. Only a few will have that privilege.

Monoculture and Overload

As much as we hope that human-centered computers may help level economic disparity across the world, the process will take time. In the shorter term, it is natural for us to wonder whether the technology might level cultural differences among the people who are interconnected. Collaboration, in the form of commerce, information work, entertainment, and education, plus individualized information access, open to the entire world the personal attitudes, customs, history, art, good and bad habits, and traits of peoples that are normally confined to citizens of single nations. Speech understanding will lead to translated exchanges that cross linguistic barriers. And automated, semantic exchanges among machines will spread shared concepts. Might these leveling forces push us toward one homogenized world culture?

When non-Americans ask this question, their dominant fear is that the answer will be "yes" and the resultant monoculture will be American. Nonsense! Tribalism is a far more powerful human force than any computing trend. Consider, for example, that although the member nations of the European Union have all been using English for a long time, it has barely affected the differences among their tribes. The Italians still differ from the British, who differ from the French, who differ from the Greeks, more or less as they have differed for centuries. What has happened among the people who participate in this sharing is the adoption of a shallow cultural layer that involves common sound bites and a few shared habits. That's exactly what I expect will happen as human-centric computing crosses national boundaries—a thin veneer of shared norms, not a monoculture.

A related fear is that the cross-border interactions will cause nations to vanish. Either their citizens will be globally distributed and won't care about national boundaries, or the ease with which the new technologies cross these boundaries will make national distinctions unnecessary. More nonsense! The police forces and armies of different nations are physically local and will remain so. They, along with their political leaders and their population, are dedicated to national survival with the same fervor that human beings are committed to personal survival. The likelihood of a military force, driven by a national political leadership, yielding its swords and bombs to some shared bits of information is pretty close to nil.

Still, the new technologies, by increasing communication, will foster a better understanding among tribes. A Greek and a Turk who love early music will join that musical "tribe" on the Net, and will get to know each other across the divide of their ancient national tribes. This could bode well for peace, since the more that people talk to one another, especially in casual settings, the less likely they are to kill their discussion partners. At the same time, these technologies will also strengthen ethnic tribes by uniting local with distributed members. For example, the 7 million Greeks living in the United States, Australia, and elsewhere outside Greece could become better tied culturally, economically, and socially with the 10 million Greeks living in the country of Hellas. Human-centric computing has the schizophrenic ability to simultaneously strengthen diversity and tribalism.

I believe that this is a great thing for our world, where these opposing forces are basic to human nature and are becoming increasingly widespread in the cities and countries where people live. The simultaneous strengthening of tribalism and diversity is yet another interpretation of how doing more by doing less might affect our world.

Another common fear is that the new technologies will overwhelm us with information, rendering us ineffective. As fashionable as this fear is, don't worry about it. Since ancient times people have valued their own survival over all else. In a serious conflict between a debilitating amount of information and survival, there is no question as to what people will do: They'll trash the information without a moment of bad conscience . . . as they should!

The Technology Fountain

As we ask the basic questions of how far we might go with human-centered computers, and how much better off we may be, we should keep in mind that technology will not stand still, and will most likely create new avenues through new discoveries. Our future vision is necessarily limited, but from what we can see, two categories of potential developments stand out—machine learning and the merger of biology with computer science. Here's why.

If computer systems become capable of learning from practice and observation of their environment, rather than by being programmed by people, we are in for a very big change. Technically, this is not part of the human-centric tool kit we have been discussing. It will require new discoveries, and as I have repeatedly said, there is no basis to predict that it will or will not happen. But if it were to succeed, we would finally have achieved great progress toward the construction of intelligent systems. Each of us would have intelligent programs and knowledgeable advisers at our side. That would bring computers even closer to serving human needs, and would result in the ultimate human-centered systems, with dramatic consequences for all of us.

Some people believe that machine learning is a dated idea and computer intelligence will evolve just as human intelligence did. They argue that since computer processing power is accelerating so much more rapidly than the human brain's, it will only take a few decades before a computer's intelligence surpasses a human's. At this point, they conclude, a machine will no longer need a human to create its offspring, and we will become irrelevant. It's fun to raise such ideas for the mental stimulation they provide. But pretending that something like this is likely to happen is quackery. What does accelerating computer power have to do with intelligence? If you move your arms faster, do you get smarter? Of course not. The growing processing power of computers says nothing about how intelligent our machines may or may not become. As for machine learning being a dated idea, discoveries are not subject to fashion like clothes! A breakthrough in machine learning, if it were to happen, would instantly become a "modern" achievement.

The evolution of machine intelligence, to where machines can beget other machines, is a metaphor that shocks and seduces, because it ascribes to future machines capabilities that people believe are uniquely human. That's even further away from our understanding than machine intelligence! People should feel free to delight in such musings. But they should not seriously worry about them any more or any less than they worry about our planet being struck by a gigantic asteroid.

The second big development that may lie ahead—a merger of biology and computer science—has nothing to do with the human-centric technologies we have been discussing. But if it were to happen, it would affect dramatically the way machines would serve us, especially for our health needs. This marriage seems plausible because biological organisms, including humans, can be characterized by their DNA structure—in other words, by information. Even though the amount of data needed to describe the molecular makeup of a single person is huge, it is still information. With the massive research effort known as the Human Genome Project as a base, scientists are increasingly able to describe in a digital blueprint the biological aspects of a person. In the imagined scenarios, this information could be used by our doctors and by us to forecast illnesses, presage hereditary strengths and weaknesses, fix or alter our human traits, and, in the extreme, to design a young fetus to have the characteristics we want it to have. In the other direction, too, biological techniques and materials could be used to fashion "computing machines" of a very different kind.

Developments like these could change the role of information in our lives, and would no doubt bring surprises. My own belief based on no facts whatsoever is that machine learning has a chance of succeeding at a partial level sometime this century. The more exotic possibility of a bio-computational merger toward the "boutique" design of living beings is too far in the future to be visible.

When we think of such possibilities, it is natural that we become frightened, to the point of asking for a moratorium on discovery, as some people have suggested, fearful of harming ourselves irreversibly with the unintended consequences of genetic engineering and machine intelligence. "Shut down the technology fountain," they say. I do not subscribe to this view, because the consequences of our discoveries are unpredictable and we are unable to chart a careful course through a universe we barely comprehend.

When we built time-shared computers and the Arpanet, we did it so we could avoid buying expensive machines, by sharing them. The efforts succeeded, not for these reasons, but because they helped people share information. The Internet was launched to interconnect networks of computers; no one expected that its biggest application would be the Web. Radar was designed for war, but ended up as a cornerstone of air transportation. Nuclear weapons research put nuclear medicine on the map. Thousands of innovations all share the same pattern—the early assessment is unrelated to the outcome. So limited is our ability to assess consequences that it's not even helped by hindsight. We can't judge whether cars, synthetic drugs, and nuclear power, all invented more than 50 years ago, are on balance good or bad for us today. Our track record of rationally assessing the future uses of science and technology is pretty lousy. How then are we going to tell what kind of research we should stop and when?

Maybe we should stop research altogether. This reminds me of a wise old airline employee. I had announced to him that I stopped flying with his company because of its poor safety record. "Listen sir," he said. "If your exit visa from this life is stamped 'death by aircraft,' even if you stay in your bed, the airplane will find you and crash upon you." At this, the dawn of the Technology Century, it is not fashionable to pay attention to forces and beliefs, like destiny, that lie outside current reason. We should reconsider. All the more so if we are arrogant enough to believe we understand our universe enough to successfully regulate its future course.

We should also remember that what we do as human beings is part of nature. I am not advocating that we do as we please, on the grounds that everything we do is natural, but rather that we respect the natural human urge to probe and understand all that surrounds us.

I suggest that as we encourage the technology fountain to feed tomorrow's discoveries and their human uses, we stay vigilant, ready to stop when danger is imminent, not when our fears or premature rational assessments, which have failed us so often, scare us into doing so. And let's ponder what other help we might seek in reaching our decisions, especially since we are not the only determinants of change out there.

As we contemplate potentially earthshaking discoveries in the context of human-centered systems, let us remember that the primary role of information in our lives is to help us achieve our human goals. Information is, therefore, a means to getting there, rather than an end in itself. That is so powerful and fundamental a property of information that together with the unchanging nature of human purpose and human beings, it is likely to survive even the wildest of tomorrow's discoveries.

No Machines beyond This Point

To fully understand the ultimate potential of human-centered computers, we should explore the limits of their uses. Is it possible that applying our new tools to certain tasks would result in our actually achieving less?

Yes. The tasks are the ones in which we convey to one another the primitive human emotions—primal forces that have been with us for thousands of years. These "forces of the cave," as I call them, range from fearing predators, seeking food and shelter, and nurturing our children to protecting our mate and trusting fellow tribe members.

By now, people who work as a team over the Internet have discovered that as long as they know and trust each other, the team functions well in its virtual forays. But when new team members join, the group loses its effectiveness. The team returns to progress only after the new members have bonded with the old ones in old-fashioned ways—by squeezing each other's hand, drinking beer together, exchanging personal stories, or giving one another a slap on the back. Building trust seems to be outside the limit of what we can do "at a distance," regardless of how faithfully the technology bridges space and time. The troubleshooter teams at British Petroleum, who use collaboration technologies to solve problems at remote oil well sites, have found this phenomenon to be true. So has MIT; as we began planning our own collaborative, distance education programs, we quickly agreed that our remote students would need to spend nearly as much time on the MIT campus as they did away from it, to partake of these deep forces that do not travel over the links of the Information Marketplace.

Why don't they? Well, imagine that your 14-year-old son has done something reprehensible. You grab him by the collar, squeeze his neck a bit, look him in the eye, and say, "Johnny, don't ever do that again." You then release your grip and explain why you were so menacing in your admonition. You could not have the same effect if Johnny were 3,000 miles away, even with the best collaboration technology that perfectly re-created your appearance, voice, and squeeze. Why? Because in the physical encounter, your son experiences a primitive fear. As you grab him, his instincts tell him that the situation could progress toward greater physical danger. Never mind if you have never struck him before. The primal forces of the cave, rather than reason, are at work. As far as these forces are concerned, there is no telling what you might do. But in the virtual scolding, your son knows, even if he is "scared" by your demeanor, that he can flip a switch and turn off the whole thing! The encounter is just a simulation. You are not transmitting primal fear to Johnny, only an image of that fear, which is no longer a primitive force.

The forces of the cave are with us all the time, regardless of the rational powers and sophisticated behaviors we invent to disguise them. And they cannot be easily tricked. Doctors healing patients, parents raising children, business associates building trust, lovers exchanging intimacy, friends accepting each other, enemies trading threats—all use the forces of the cave. Even though the information component of these activities could be communicated well with human-centered machines, the exchanges would be nowhere as effective.

The forces of the cave set a clear limit as to how far human-centric computing can go toward helping us do more by doing less. Even when we finish the Unfinished Revolution, they will still hold sway.

Greater Humanity?

We want to go beyond the efficiency, ease of use, fun, and productivity implications and explore whether the human-centric technologies can "do more" to enhance our humanity, to truly make us "better off." To ponder this lofty question, we must declare what we consider being "human" signifies. Each of us assigns a highly individual interpretation to this term, since it defines the meaning and purpose of our unique lives.

We can't get a universal definition, but we can characterize several of the dimensions that constitute what humanity might mean to different people. Then we can assess how the new technologies may or may not help us along each dimension. By selecting which of the dimensions you deem important, you can get an idea as to how human-centric computing might affect your own sense of being human.

During the Enlightenment, people decided to separate reason from faith and from the literature of the ancients. This freed science and technology from the shackles of religion and humanism. It fueled the Industrial Revolution and later the Information Revolution. The success of industrialization confirmed the wisdom of separating these dimensions of humanity, and reinforced the three-way separation among technologists, who put their faith in reason; humanists, with their focus on the arts, literature, and human feelings; and believers centered on spirituality.

Here, then, are three historically vetted dimensions of what it might mean to be human: the *reason* part that stands behind science, technology, and rational thinking; the *feeling* part that lifts the arts and the humanities; and the *faith* part that helps us cope with what cannot be explained or felt. Add our physical *action* and we cover a good deal of what it means to be human. Where do you fall along these dimensions? Which do you consider more important in your own life? As you formulate your answer, let's take a look at whether, or how, the human-centric technologies might affect each dimension.

The rational part of being human will benefit greatly, because it is the stuff of which the technologies are made. We have seen many ways to enhance our reason through greater access to information, better communication, customization, and much more.

Automation can amplify the action part of our humanity by bringing the physical world under our greater control, and by harnessing our machines to act in our stead. Planning, crucial to future action, is also dramatically enhanced by having access to good information and being able to process and share it effectively.

How about the feeling dimension? We have just established that the primitive forces are outside the reach of the new technologies. However, that doesn't mean that emotions can't be conveyed by the virtual world. We all laugh and cry at a good story or movie that reaches us over the Net, so certainly lighter-than-primal emotions can be sustained. We can intensify sensory perception, too, by brightening colors and sounds, and perceiving sensations across greater distance. We can read more, access a great deal of the world's art, and use aids that help us when we create a poem or a picture. But we can't emote more deeply through the new technologies. The audience of a large-screen, 3-D, multimedia video packed with visual and sound effects cannot be made to feel more sincere empathy with the victims of a plane crash than you do when you read a good, plain-text newspaper article about the tragedy. The new technologies can amplify the feeling part of our humanity in a quantitative and somewhat perfunctory sense, but they cannot make us feel more deeply.

That brings us to faith. It is hard to imagine how a person's spirituality could be enhanced by technologies that deal with information. After all, faith, in those that have it, is essentially defined as something internal to our being and outside the realm of human reason, feeling, or action. The new human-centric technologies cannot amplify the spiritual dimension of our humanity.

If you are a hard-core technologist who believes that rationality is the essence of being human, or if you are a driven person who believes in action, then human-centric computing will greatly enhance your humanity. If you are an artist who thrives on feelings and new ways of expressing the world, you will find partial enhancement from the new technologies. If you are a monk whose life revolves around spirituality, you will look elsewhere for help. But if you possess varying amounts of these human dimensions—which describes most of us—then you can determine how much "better off" your humanity will be by analyzing how each of the dimensions you care about will be affected.

Beyond the Information Revolution

To my thinking, the ultimate way in which we can do more by doing less goes beyond the Information Revolution but is made all the more urgent by its growing dominance. It involves the way we reconcile the human dimensions within us.

The millennium that just ended was dominated by God and faith, reflected in religious wars from the Crusades to the ongoing Middle East crisis, the split from Orthodoxy, the Reformation, and centuries of music and art that stemmed overwhelmingly from religion. Now, as the new millennium begins, this dominance is shifting toward a new "god"—technology—which began its powerful ascent toward the end of the 20th century. People stand awestruck by the miracles of information technology, biotechnology, medicine, and materials science, which promise to transform our behavior, our being, and our surroundings. They increasingly place their faith in this new god to address their human needs for better health, protection from danger, explanation of our surrounding world, and greater happiness. Since technology, and especially information technology, thrives on reason, the new millennium of technology, left unchecked, will further enhance reason at the expense of feeling and faith, aggravating the separation among these three pieces of our humanity.

That separation grew as the Industrial Revolution became increasingly successful, and led to problems. Technologists began questioning their purpose. Humanists became disaffected with gadgets and materialistic ideas. The spiritually inclined resented the loss of beliefs. Youth, sensing that something was missing inside them, turned to apathy and drugs. People focused increasingly on themselves, celebrating possessions, lamenting depressions, and fragmenting families. Governments separated faith from reason in the school curricula. A politically correct population became increasingly reluctant to say "God." Universities isolated technologists from humanists in watertight compartments across campus from each other. Today the separation has become so ingrained we don't even see it or the problems it has engendered. We simply accept it as "natural."

If we allow this trend to continue, our problems will increase and we will miss the prospect of being better off in the biggest possible sense of being human. We simply can't go far if we stay fragmented. Take humanism; until recently, the essence of being well educated was, in the words of the English poet Matthew Arnold, "to know the best that has been thought and said in the world." If you needed technology you bought it, like potatoes, to serve your loftier humanistic goals. That's how technologists became known as practitioners of "the servile arts." This humanist-dominant view made sense when technology was a small part of our lives—a notion that is no longer valid! Today, higher purpose may *originate* with technology, as in the invention of the Web by a full-fledged technologist. Many sites with a purely social purpose, developed by technologists, are already in operation. No pure humanist could ever have come up with these ideas, without also understanding technology. It's time for Matthew Arnold's words to be qualified. Technology will be as important a contributor to noble endeavors and understanding our world as humanistic ideals were and will continue to be. Keeping the technologists separated from the humanists will keep us from discovering these new territories.

People also have an inherent need for spirituality, which offsets the powerlessness we feel before the many mysteries that surround us. In an increasingly rational world, how might our children fulfill this human need, which has led billions to religion throughout the centuries? Never mind grandstanding on the industrial world's easy answer that church and state should stay separated, and the latter shouldn't glorify any particular sect in the schools. Good. Let's keep doing that. But then what? Will learning in the next millennium stay chained to reading, arithmetic, and reason? What of birth, friendship, love, marriage, illness, divorce, conflict, death, purpose?

If we remain fragmented, we'll be unable to fulfill our full human potential, because we will be running on only some of our cylinders. People lived for thousands of years without this internal separation. And we were not always as impressed with reason, morality, and all that we have built on the shaky foundation of human thought as we have been in the last few centuries. It is ironic, yet inescapable, that so many "thinkers," especially Western philosophers, stayed chained to reason and built their theories upon it, as if it were the only solid ground. Granted, we can't help but be impressed by this unique capability of our brain, which in its exquisite architecture and processes holds our awesome power to think. Yet, viewed from afar, it is just another property of a few ounces of meat tucked inside the skulls of antlike creatures that roam a huge earth in an infinite universe. What does reason have to do with the love of a child, the beauty of a flower, the eternity of stone, our origin, our destination? The new century of technology is amplifying our tendency to overrate reason at the expense of spirituality, and technological reason at the expense of humanistic ideas.

Do you find such philosophical considerations too abstract? Do you prefer to stay practical? Then here's something for you: How do we cope with children who use guns to kill their classmates? What do we do when genetic engineering can alter the personality of a fetus? How do we deal with trans-border crimes over the Internet? And how about all the other "ordinary" problems we will face that won't be as famous as these, but just as hard? Every decision we make, whether it's choosing a school for our children, managing people, cementing or breaking relationships, facing illness, running a household or a company or a country, will increasingly involve issues and considerations that are intertwined across these artificial divisions. Pure technology can't solve these problems. Nor can pure humanism or pure faith. We need to bring these back together if we want to find our way through the maze of an increasingly complex world.

This is especially true as we begin our journey to finish the Unfinished Revolution. The human-centric technologies will bring computers closer to us and give us power to do more by doing less. But the highest meaning of "human-centric," and its biggest benefit to us, will be determined by what we do to achieve the human goals we set. We will be better off and we will be finishing the ultimate Unfinished Revolution if we reach for these goals using all our human dimensions in concert, standing once again in awe before the sunset, the wheel, and what may lie behind them.

Index